27·50 ✓

41556

C4238

TIG and PLASMA
Welding

Process techniques, recommended practices and applications

W Lucas, DSc, PhD, CEng, FIM, FWeldI

Published in association with
Huntingdon Fusion Techniques

ABINGTON PUBLISHING

Woodhead Publishing Ltd in association with The Welding Institute
Cambridge   England

Published by Abington Publishing
Woodhead Publishing Ltd, Abington Hall, Abington,
Cambridge CB1 6AH, England

First published 1990

British Library Cataloguing in Publication Data
Lucas, W
TIG and plasma welding
1. Plasma welding and tungsten inert gas welding
I. Title II. Series
671.5212

ISBN 1 85573 005 7

Typeset by Goodfellow & Egan, Cambridge and
printed by Crampton & Sons Ltd, Sawston, Cambridge

Foreword

With ever increasing attention to quality control and automation in welding, TIG and plasma processes are becoming more widely used.

Huntingdon Fusion Techniques Limited has specialised in providing precision welding equipment and accessories since 1975 and are delighted to be associated with this informative book.

Dr Lucas touches on most aspects of TIG and plasma welding which should give this work a wide appeal to all people wishing to use a high quality controlled heat input welding process.

We commend 'TIG and Plasma Welding' to all people involved with teaching, design and manufacturing as an easy to read and comprehensive book, which should not date.

Ron A Sewell
Managing Director
Huntingdon Fusion Techniques Ltd

SPECIALISTS FOR 15 YEARS IN
TIG & PLASMA AUTOMATIC WELDING

Huntingdon Fusion Techniques Limited is a specialist welding company manufacturing equipment for industries such as: nuclear, petrochemical, aerospace and general industrial plants. Here are some of our quality products.

Our **video arc monitor** (right) allows operators to view a very highly magnified, colour image of electrode, weld joint fit-up, cleanliness etc. It is a valuable inspection and training aid and is useful for weld procedure development.

State-of-the-art **TIG and plasma power supplies** provide precise setting of welding parameters which are accurately reproduced weld after weld.

Orbital welding systems with complete programming consistently produce highly reliable tube welds. A range of **welding heads** allow "push button" completely automatic welding of tubes 1.5-250 mm diameter to meet demanding welding criteria.

When TIG and plasma welding require the addition of wire to the weld pool, we provide the products to simplify the process. Our range of **wire feeders** gives you choices for manual and mechanised feeding of cold and hot wire and our **arc voltage controller** provides continual weld gap regulation on parts of irregular contour.

Multi-strike tungsten electrodes contain a new material which greatly enhances electrode life by increasing the number of strikes per tungsten. They do not contain any radioactive material and operate at lower temperatures giving cooler welds.

To prepare tungstens, our range of **tungsten electrode grinders** are ideal for manual or automatic welding techniques. They provide consistent and repeatable points which greatly enhance arc ignition and stability, promoting a smooth electron flow from the tungsten to the work. They provide rapid payback by extending the life of your tungsten and time between re-grinds.

A range of argon **purging chambers** (left) for the welding of titanium and other reactive metals under inert gas conditions. As well as standard or adapted chambers, we also make special purge chambers to order.

We also manufacture systems, offering standard and tailor-made designs.

FOR FURTHER DETAILS, CONTACT . . .

HUNTINGDON FUSION TECHNIQUES LIMITED
STUKELEY MEADOWS INDUSTRIAL ESTATE
HUNTINGDON   CAMBRIDGESHIRE   PE18 6ED

TELEPHONE (0480) 412432
TELEX 32151 HFTLTD G
FAX (0480) 412841

HUNTINGDON FUSION TECHNIQUES

CONTENTS

Preface   7

Part I TIG WELDING

Chapter 1   PROCESS FUNDAMENTALS   9
               DC TIG   9
               Pulsed current   18
               AC TIG   22

Chapter 2   APPLYING THE TIG PROCESS   25
               Practical considerations   25
               Manual welding   29
               Mechanised operation   37
               Typical defects   40

Chapter 3   MECHANISED ORBITAL TUBE WELDING   46
               Welding techniques   46
               Equipment   47
               Applications   50

Chapter 4   TUBE TO TUBEPLATE WELDING   56
               Welding techniques   56
               Equipment   56
               Applications   57

Chapter 5   MICRO-TIG WELDING   62
               Welding techniques   62
               Equipment   63
               Applications   63

Chapter 6   HOT WIRE TIG WELDING   65
               Welding techniques   65
               Equipment   65
               Applications   65

Chapter 7   NARROW GAP TIG WELDING   73
               Welding techniques   73
               Equipment   73
               Applications   76

Part II PLASMA WELDING

Chapter 8	PROCESS FUNDAMENTALS	80
	DC Plasma welding	80
	AC Plasma welding	83
	Pulsed current (keyhole) welding	85
Chapter 9	APPLYING THE PLASMA PROCESS	90
	Practical considerations	90
	Industrial applications	97
Chapter 10	THE FUTURE	109
Index		110

Preface

The tungsten arc welding processes are currently exploited widely for precision joining of crucial components and those which require controlled heat input. The small intense heat source provided by the tungsten arc is ideally suited to the controlled melting of the material. As the electrode is not consumed during welding, as with the MIG (GMA) or MMA (SMA) welding processes, autogenous welding can be practised without the need for continual compromise between the heat input from the arc and the deposition of the filler metal. Because the filler metal, when required, can be added directly to the weld pool from a separate wire feed system, all aspects of the process can be precisely and independently controlled, i.e. the degree of melting of the parent metal is determined by the welding current with respect to the welding speed, whilst the degree of weld bead reinforcement is determined by the rate at which the filler wire is added to the weld pool.

Within the context of gas tungsten arc welding two quite distinct processes have emerged - TIG and plasma welding. Whilst both are equally suitable for manual and mechanised welding, certain operating modes can be exploited for specific applications. These unique modes are derived almost exclusively from the electrode/torch configuration and the gas flow system. However, within the two welding processes there are a number of important operating techniques or process variants which can almost be considered as welding processes in their own right. The variants of interest include:

Pulsed current (TIG and plasma);
Micro-TIG;
TIG-hot wire;
Narrow gap TIG;
Keyhole plasma.

To aid a full understanding of the operating features of the TIG and plasma processes and their variants, i.e. with a view to exploiting the advantageous features, information is initially presented on the fundamental electrical, arc and process characteristics. However, because of the similarities in their operating characteristics, the welding engineer often has to make a difficult choice between techniques. Practical experience gained at The Welding Institute in evaluating the techniques, together with the production experience of its Research Members, have shown that there are areas where the special features of each technique offer specific advantages. Thus, considerable emphasis is placed on describing current applications including operating data, in a wide range of components from the various sectors of industry. It is hoped that by elucidating the reasons for the choice of a particular technique, readers will be better placed to make the best use of TIG and plasma welding in their own company.

The author acknowledges process research and development data and helpful discussions on the technical aspects of TIG and plasma welding with colleagues at The Welding Institute, in particular J C Needham, G A Hutt, M R Rodwell, I D Harris and M D F Harvey. Particular thanks are extended to D Patten, B O Males and M G Murch for guidance on the practical techniques described and the information provided on the application of the TIG and plasma welding processes.

Help with the preparation of the manuscript and the drawings by Mrs J M Lucas, M J Lucas and W B Lucas is gratefully acknowledged. Abington Publishing would also like to thank Mr R Sewell for his helpful advice.

The following companies have kindly agreed to the publication of technical information and photographs of applications: Air Products plc, British Alcan Ltd, Babcock Energy Ltd, Foster Wheeler Power Products Ltd, ESAB, Huntingdon Fusion Techniques Ltd, Precision Systems Ltd, APV Paramount Ltd, Skomark Engineering Ltd, Devtec Ltd, William Press Ltd, Darchem Engineering Ltd, SAF Welding Products Ltd, Abia, BOC Ltd, Arc Machines Inc, Saipem SpA, Equipos Nucleares SA, Kobe Steel Company, Sciaky Bros Ltd, Rolls Royce Plc (Rodney Fabrication Facility), Jet Joint Undertaking.

CHAPTER 1
Process fundamentals

DC TIG

In tungsten inert gas (TIG or GTA) welding the arc is formed between a pointed tungsten electrode and the workpiece in an atmosphere of argon or helium, Fig. 1. In DC welding, the electrode usually has negative polarity (US nomenclature is DC straight polarity) - its electron thermionic emission properties reduce the risk of overheating which may otherwise occur with electrode positive polarity (Fig. 1a). The ionised gas or plasma stream thus formed can attain a temperature of several thousand degrees centigrade, at least in the central core of the arc near to the electrode. Consequently, within the normal range of welding currents from a fraction of an ampere to several hundred amperes (selected according to the thickness of the material) rapid melting can be effected. However, the operation of tungsten arc processes in practice is essentially very simple as the heat required to melt the metal is determined merely by setting the welding current relative to the welding speed, generally within the range 0.1-300A.

As the electrode is not consumed during welding, additional metal e.g. when required to fill a joint, must be added separately in the form of a wire rod (Fig. 1b).

The gas supplied to the arc has two functions; it generates the arc plasma, and it protects the electrode weld pool and weld bead from undesirable oxidation. The arc is in the form of a cone (Fig. 1b), the size of which is determined by the current, the electrode diameter and vertex angle, but the penetration characteristics are primarily determined by the current level and the shielding gas composition.

Electrode

Selection of electrode composition and size is not completely independent and must be considered in relation to the operating mode and the current level. Electrodes for DC welding are pure tungsten or tungsten with 1, 2 or 4% thoria, the thoria being added to improve electron emission which

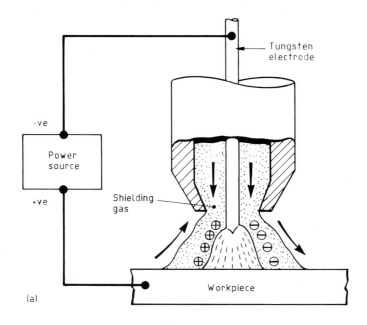

Tungsten
electrode

-ve

Power
source

+ve

Shielding
gas

Workpiece

(a)

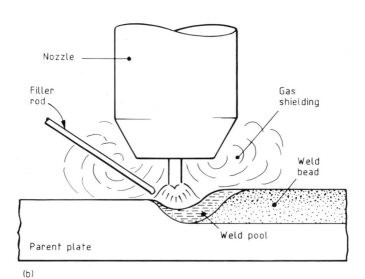

Nozzle

Filler
rod

Gas
shielding

Weld
bead

Weld pool

Parent plate

(b)

10

1 *The TIG welding process: a) Torch and power source arrangement; b) TIG welding operation; c) Characteristic appearance of the DC TIG arc and weld pool; d) Characteristic appearance of the AC TIG arc showing cathodic cleaning of the plate surface.*

facilitates arc ignition. Alternative additions to lower the electron work function are lanthanum oxide or cerium oxide, which are claimed to improve starting characteristics, provide excellent arc stability, lower electrode consumption and replace thoria which is radioactive. When using thoriated electrodes it is recommended that precautions are taken in their handling and storage and if possible avoid contact with grinding dust and smoke. In DC welding, a small diameter, finely pointed (approximately 30°) electrode must be used to stabilise low current arcs at less than 20A. As the current is increased, it is equally important to readjust the electrode diameter and vertex angle. Too fine an electrode tip causes excessive

Table 1 Recommended electrode diameter and vertex angle for TIG (GTA) welding at various current levels

| Welding current | DC, electrode negative | | | AC | |
| | Electrode* diameter | | Vertex angle | Electrode† diameter | |
A	mm	in	degrees	mm	in
<20	1.0	0.040	30	1.0–1.6	0.040–$^1/_{16}$
20–100	1.6	$^1/_{16}$	30–60	1.6–2.4	$^1/_{16}$–$^3/_{32}$
100–200	2.4	$^3/_{32}$	60–90	2.4–4.0	$^3/_{32}$–$^5/_{32}$
200–300‡	3.2	$^1/_8$	90–120	4.0–4.8	$^5/_{32}$–$^3/_{16}$
300–400‡	3.2	$^1/_8$	120	4.8–6.4	$^3/_{16}$–$^1/_4$

* Thoriated tungsten

† Zirconiated tungsten, balled tip, electrode diameter depends on degree of balance on AC waveform; for balanced waveform use larger diameter electrode

‡ Use current slope-in to minimise thermal shock which may cause splitting of the electrode

broadening of the plasma stream, due to high current density, which may result in a significant decrease in the depth to width ratio of the weld pool. More extreme current levels result in excessively high erosion rates and eventually in melting of the electrode tip. Recommended electrode diameters and vertex angles in argon shielding gases for the normal range of currents are given in Table 1.

In AC welding, where the electrode must operate at a higher temperature, the positive half-cycle generates proportionally more heat in the electrode than when operating with electrode negative polarity. A pure tungsten or tungsten-zirconia electrode is preferred, as the rate of tungsten loss is somewhat less than with thoriated electrodes. Furthermore, because of the greater heating of the electrode, it is difficult to maintain a pointed tip and the end of the electrode assumes a spherical or 'balled' profile (Fig. 1d).

Power source

The power source necessary to maintain the TIG arc has a drooping voltage-current characteristic which provides an essentially constant current output even when the arc length is varied over several millimetres (Fig. 2a). Hence, the natural variations in arc length which occur in manual welding have little effect on welding current level. The capacity to limit the current to the set value is equally crucial when the electrode is short circuited on to the workpiece. Otherwise, excessively high currents are drawn, damaging the electrode and even fusing the electrode to the workpiece.

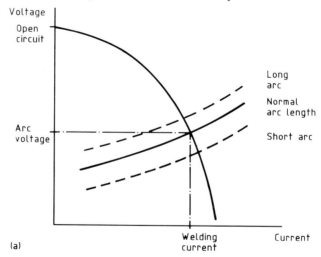

(a)

12

2 *Power sources for TIG welding: a) Operating characteristics of a constant current power source; b) Inverter power source for manual welding (courtesy of ESAB); c) Transistor (100A) power source for mechanised welding of precision components (courtesy of Huntingdon Fusion Techniques Ltd).*

In practical operation the power source is required to reduce the high voltage mains supply, 240 or 440V, AC, to a relatively low open circuit voltage, 60-80V, AC or DC. In its basic form, the power source comprises a transformer to reduce the mains voltage and increase the current, and a rectifier, placed on the secondary side of the transformer to provide the DC supply. Traditional power source designs use a variable reactor, moving coil or moving iron transformers, or a magnetic amplifier to control the welding current. Such equipment has the highly desirable features of simple operation and robustness, making it ideally suited to application in aggressive industrial environments. The disadvantages are relatively high material costs, large size, limited accuracy and slow response. More recently, electronic power sources have become available which do not suffer from these disadvantages, but at their present stage of development are more expensive. The various types of electronic power source are:

– Thyristor (SCR), phase control;

– Transistor, series regulator;

– Transistor, switched;

– AC line rectifier plus inverter.

The major operating features of these systems, with their advantages and disadvantages, are given in Table 2. Of the power source designs listed, the transistor based, series regulator control systems offer greater accuracy and reproducibility of welding parameters, but tend to be wasteful of electrical energy. The AC line rectifier plus inverter type offers the combination of high electrical efficiency and small size. Examples of compact commercially available power sources for manual and mechanised operation are shown in Fig. 2b and c.

Shielding gas

The shielding gas composition is selected according to the material being welded, and the normal stage of commercially available gases is given in Table 3. In selecting a shielding gas it should be noted that:

1 The most common shielding gas is argon. This can be used for welding a wide range of materials including mild steel, stainless steel, and the reactive metals—aluminium, titanium and magnesium.

2 Argon-hydrogen mixtures, typically 2% and 5%H_2, can be used for welding austenitic stainless steel and some nickel alloys. The advantages

14

Table 2 Major operational features of electronic power sources compared with conventional variable reactor or magnetic amplifier power sources

Control type	Method of control	Advantages	Disadvantages
1 Thyristor (SCR) phase	SCRs replace diodes on secondary output of the transformer Alternatively, triacs or inverse parallel SCRs used in the primary of the transformer	1 Better accuracy of current and time settings 2 Can be used to produce square wave AC waveform 3 Can be used for pulsed operation	1 High ripple unless large amount of inductance is placed in series with output 2 Pulsed response normally limited to 100Hz
2 Transistor series regulator	Power transistors in parallel, analogue control from low current input signal	1 Very stable and accurate control of current level – better than 1% of set level 2 Pulsing over wide range of frequencies, up to 10 kHz, and pulse shape can be varied	1 Poor electrical efficiency 2 DC supply only
3 Transformer with secondary (transistor) chopper	Transistor, high frequency switching of DC supply	1 Accuracy and control similar to series controller 2 Less wasteful of energy compared with series controller 3 Greater arc stiffness can be exploited for low current operation	1 Although similar output to series controller, pulse frequency and wave shaping less flexible
4 AC line rectifier plus inverter	Mains supply rectified to high voltage DC then converted by transistors or SCRs to AC operating at 2–20 kHz. Final output produced by small mains transformer and rectified to DC	1 Because transformer operates at high frequency, the size and weight of the mains transformer can be greatly reduced 2 Because of its small size, cost of raw materials significantly reduced 3 High electrical efficiency and high power factor	1 Response rate not as high as transistor controlled power source

Table 3 Recommended shielding gases for TIG welding

Metal	Shielding gas mixtures					
	Argon	Argon +H$_2$	Helium	Helium-argon	Nitrogen	Argon-nitrogen
Mild steel	●					
Carbon steel	●			○		
Low alloy steel	●			●		
Stainless steel	●	●	○	○		
Aluminium	●		●	●		
Copper	●		●	●	○	○
Nickel alloys	○	●		○		
Titanium and magnesium	●		○			

● most common gas
○ also used

of adding hydrogen are that the shielding gas is slightly reducing, producing cleaner welds, and the arc itself is more constricted, thus enabling higher speeds to be achieved and/or producing an improved weld bead penetration profile, i.e. greater depth to width ratio. It should be noted that the use of a hydrogen addition introduces the risk of hydrogen cracking (carbon and alloy steels) and weld metal porosity (ferritic steels, aluminium and copper), particularly in multipass welds.

3   Helium and helium-argon mixtures, typically 75/25 helium/argon, have particular advantages with regard to higher heat input; the greater heat input is caused by the higher ionisation potential of helium which is approximately 25eV compared with 16eV for argon. As the helium based shielding is considerably 'hotter' than an argon based gas, it often promotes higher welding speeds and improves the weld bead penetration profile. The reluctance to exploit the benefits of helium is directly associated with its cost, which in special gas mixtures may be as much as three times that of argon. A secondary disadvantage in employing helium rich gases is the difficulty often experienced in initiating the arc, which can be particularly severe in pure helium.

4   As nitrogen is a diatomic gas, on re-association at the workpiece surface, it is capable of transferring more energy than monatomic argon or helium. Hence its addition to argon can be particularly beneficial when welding materials such as copper, which have high thermal conductivity; the advantages of nitrogen additions cannot be exploited when welding

ferritic and stainless steels, because nitrogen pick-up in the weld pool would cause a significant reduction in toughness and corrosion resistance.

Flow rate setting

Because of the inherent risk of porosity in TIG welding, the importance of efficient gas shielding cannot be stressed too highly. The shielding gas flow rate is influenced by the following factors:

– Nozzle diameter;

– Current level;

– Type of current;

– Electrode stickout;

– Shielding gas composition;

– Type of joint;

– Welding position.

Recommended shielding gas flow rates for various practical situations are given in Fig. 3. It should be noted that the flow rate should be increased

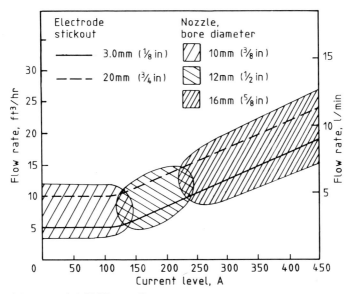

3 Recommended shielding gas flow rates.

17

when the electrode stickout, welding current or nozzle diameter is increased. In AC operation, as the arc is generally broader and the current reversals have a greater disturbance effect on the shield the flow rates should be increased by approximately 25% compared to DC welding at the same current level.

The effectiveness of a gas shield is determined at least in part by the gas density. As the density of helium is approximately one tenth that of argon, difficulties can be experienced in protecting the weld pool, particularly when welding under draughty conditions or at high currents, which may induce turbulence in the gas shielding stream. However, effective shielding can be maintained by increasing the gas flow, typically by a factor of two.

Shielding of the weld pool area can also be improved by the use of a gas lens, which is inserted into the torch nozzle to ensure laminar flow. Adoption of this technique is strongly recommended when welding in positions other than flat and for welding curved surfaces. When welding corner or edge joints, excessive flow rates can cause the gas stream to bifurcate which may result in air entrainment. The effectiveness of the gas shield can often be improved by reducing the gas flow by approximately 25% and here the use of a gas lens is considered essential.

Pulsed current

Principles

Particular mention must be made of the pulsed current technique as applied to TIG welding. The essential feature is that a high current pulse is applied causing rapid penetration of the material. If this high current were maintained, excessive penetration and ultimately burn-through would occur. Therefore, the pulse is terminated after a preset time and the weld pool is allowed to solidify under a low background or pilot arc. Thus the weld progresses in a series of discrete steps with the pulse frequency balanced to the traverse rate to give approximately 60% overlap of the weld spots. The surface appearance of a typical pulsed current weld is shown in Fig. 4a.

The pulsed technique has been found to be particularly beneficial in controlling penetration of the weld bead, even with extreme variation in heat sink. Such variations are experienced either through component design, thick-to thin sections, or from normal production variations in component dimensions, fit-up, clamping and heat build-up. In conventional continuous current welding, where a balance must always be achieved between the heat input from the arc, the melting to form the weld pool and the heat sink

18

4 *Pulsed current operation: a) Surface
appearance of a pulsed TIG weld showing
how welding progresses in a series of
overlapping spot welds; b) Pulsed TIG
weld between 0.1mm thick, AISI 321
stainless steel and 1mm thick mild steel;
major pulse parameters, pulse current 75A,
pulse time 70ms, background current 15A,
background time 140ms; c) Pulsing used to
control root penetration when welding pipe
in the fixed horizontal (5G) position.*

represented by the material or component being welded, penetration is greatly influenced by these variations. However, in pulsed operation, rapid penetration of the weld pool during the high current pulse and solidification of the weld pool between pulses markedly reduce the sensitivity to process variation through the effects of heat build-up and/or disparity in heat sink.

An example of pulsed TIG welding between thin wall convoluted stainless steel tube and relatively thick wall mild steel tube is shown in Fig. 4b. Pulsing is also used to control root penetration when welding pipe in the fixed horizontal (5G) position, as shown Fig. 4c.

Parameter settings

Despite the obvious advantages of the pulsed process in production, the technique may appear to be a further complication in that a greater number of welding parameters must be considered, i.e.:

- Pulse time;

- Pulse level;

- Background time;

- Background level.

The technique can be simplified in the first instance from the knowledge that, for a given material, there is a preferred pulse level which is based on its diffusivity and, to a lesser extent, on its thickness. The preferred currents are approximately 300A for copper, 150A for carbon steel, 100A for cupronickel and 50A for lead. Thus, for a given component, the operator need only set the pulse time to achieve penetration which is determined solely by thickness. For example, for welding 2.5mm (0.1in) stainless steel at 100A, a 0.4sec pulse would be demanded, whilst for a 1.5mm (0.06in) thick material, the pulse time would be reduced to 0.1sec at the same current level. The background parameters are considerably less critical in the pulsing operation. The background level is normally set at approximately 15A, which provides the greatest possible heat dissipation during this period whilst being high enough to maintain a stable arc. The background period is normally equal to the pulse period but may be some two or three times greater in welding thicker sections.

This approach is presented only as a guideline for the initial selection of welding parameters, and must be treated with caution, particularly when welding at the extremes of the thickness range, i.e. sections of greater than 3.0mm (0.12in) and less than 1mm (0.04in). In both instances, the preferred pulse current level will be outside the above theoretical operating ranges. For example in welding stainless steel, practical trials have established that for a thickness of 4mm (0.16in) the preferred pulse parameters are 200A/0.75sec, whilst for 0.5mm (0.02in) thick material, the preferred pulse parameters are 50A/0.1sec (Fig. 5).

Welding thick sections at too low a pulsed current can result in loss of most of the advantages of pulsing (controlled depth of penetration and tolerance to variation in heat sink) as the weld pool takes a long time to penetrate the material and thermal diffusion occurs ahead of the fusion front. In welding thinner sections with too high a pulsed current, the excessive arc forces may cause cutting and splashing of the weld pool, resulting in a poor bead profile and electrode contamination.

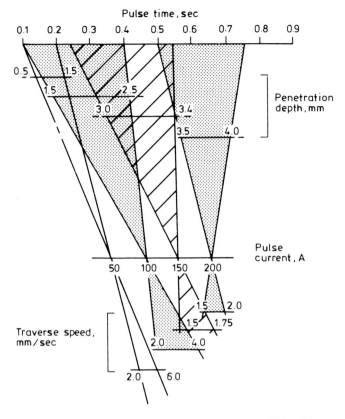

5 *Nomograph as an aid to the selection of pulse parameters in TIG welding.*

The capacity to use lower pulsed currents and longer pulse times is also of particular importance when using power sources which have a limited response, i.e. a low rate of rise and fall of the current between the background and the pulsed current levels. For instance, power sources in which the current is controlled by a magnetic amplifier are generally limited to pulses of 0.2sec duration, whilst in thyristor controlled types the response is markedly improved and pulses as short as 0.03sec can be generated. However, for complete flexibility, transistor controlled power sources are used which can generate pulses within an almost unlimited frequency range up to 10kHz. An added advantage of these power sources is the capacity to reproduce accurately complex pulse waveshapes which can be of benefit in controlling the weld pool and solidification structure.

21

AC TIG

Sine wave arc

TIG welding is also practised with AC, the electrode polarity oscillating at a frequency of 50Hz. The technique is used in welding aluminium and magnesium alloys, where the periods of electrode positive ensure efficient cathodic cleaning of the tenacious oxide film on the surface of the material (Fig. 1d). Compared to DC welding, the disadvantages of the technique lie in the low penetration capacity of the arc and, as the arc extinguishes at each current reversal, in the necessity for a high open circuit voltage, typically 100V and above, or continuously applied HF, to stabilise the arc. Low penetration results in particular from the blunt or 'balled' electrode which is caused by the high degree of electrode heating during the positive half-cycle. Where deep penetration is required, use of DC with helium as the shielding gas, which does not suffer from these disadvantages and is somewhat tolerant to surface oxide, may be an alternative. Use of helium, however, is not particularly attractive because of its high cost and, in the absence of the cleaning action of the arc, the weld pool/parent metal boundaries can be somewhat indistinct, thus making it difficult to monitor and control the behaviour of the weld pool.

Square wave arc

A new generation of AC power sources has recently become available; their principal feature is that the output current assumes a more square waveform, compared with the conventional sine wave (Fig. 6). Two types of power source are available, differing in the manner in which the square waveform is produced. Whilst a 'squared' sine waveform is generated by using inverted AC, a more truly square waveform is produced by a switched DC supply.

In either case the importance for TIG welding is that the current is held relatively high prior to zero and then transfers rapidly to the opposite polarity. In comparison, the current developed by sine wave power sources decreases more slowly to current zero and likewise the current built up after re-ignition is at a much lower rate.

The benefit of square wave AC is that, aided by the inherent high surge voltage associated with the rapid current reversal, AC TIG can in some instances be practised at 75V without the need for HF spark injection to be superimposed for arc re-ignition.

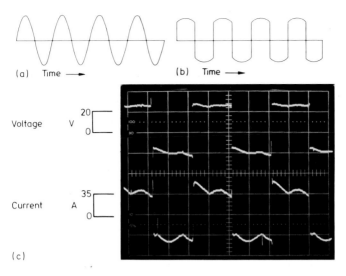

6 *Characteristic re-ignition waveforms for: a) Sine wave supply at 100 OCV; b) Switched DC supply at 75 OCV; c) Square wave AC power source.*

An additional feature of square wave AC power sources is the capacity to imbalance the current waveform, i.e. to vary the proportion of electrode positive to electrode negative polarity. In practice, the percentage of electrode positive polarity can be varied from 30-70% at a fixed repeat frequency of 50Hz. By operating with a greater proportion of electrode negative polarity, heating of the electrode can be substantially reduced compared with that experienced with a balanced waveform. Although cleaning of the oxide on the surface of the material is normally sufficient with 30% electrode positive, the degree of arc cleaning may be increased by operating with a higher proportion of electrode positive (up to a limit of approximately 70%).

Further reading

1 Wiles L J 'Shielding gas mixtures as used in MIG and TIG welding'. *Australian Weld J* 1977 21 2.

2 Metcalfe J C and Quigley M B C 'Arc and pool stability in GTA welding'. *Weld J* 1977 55 5.

3 Giedt W H, Tallerico L N and Fuerschibach: 'GTS welding efficiency, calorimetric and temperature field measurements'. *Weld J* 1989 68 1.

4 Lu M and Kou S: 'Power and current distributions in gas tungsten arcs'. *Weld J* 1988 67 2.

5   Yoryachev A P and Zelenin V A 'Mechanical deep penetration arc welding with a tungsten electrode'. *Automatic Welding* 1964 12.

6   Needham J C 'Joining metals by the pulsed TIG process - a solution to many welding problems'. *Australian Weld J* 1972 16 5.

7   Grist F J 'Improved, lower cost aluminium welding with solid state power source'. *Weld J* 1975 54 5.

8   Yamamoto H: 'Recent advances in arc welding equipment and their application'. Welding Technology Japan, The Welding Institute, 1984.

CHAPTER 2

Applying the TIG (GTA) process

Practical considerations

The TIG process is used extensively in all branches of industry e.g. chemical and nuclear plant and aero-engineering industries. The principal type of application is one in which quality is paramount such as welding thin material down to 0.5mm (0.02in) thickness, and for precision welding heavier components. The relatively small arc can be precisely positioned on the joint and heat controlled to minimise distortion. In butt welding of material within the thickness range 0.5–3mm (0.02–0.12in), welding is normally carried out autogenously, i.e. without the addition of filler material. However, if the joint configuration contains a gap, weld bead reinforcement is required, or if the material is sensitive to weld metal cracking or porosity, filler must be applied. It can be added in wire or rod form with composition either of a matching analysis, or of a specific composition to overcome a metallurgical problem, e.g. containing deoxidants to prevent formation of porosity in the weld metal.

Joint preparations

Sheets of thickness less than 3mm (0.12in) are welded with a simple square edge butt joint configuration, see Fig. 7a and b. Sheet and plate thicker than 3mm (0.12in) require an edge preparation and typical joint configurations are given in Fig. 7c and d. The root may be completed autogenously, provided that there are no metallurgical problems, and the joint is then completed with the required number of passes with filler added to the weld pool.

Alternative joint preparations are available for tubular components and these are discussed separately in Chapter 3.

Backing systems

For fully fused welds, sheets should be clamped to a rigid, temporary, backing bar which supports the penetration bead during welding. Use of a

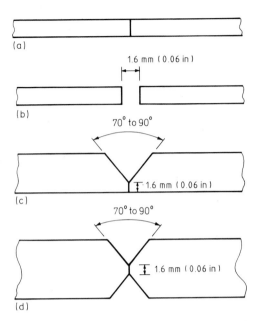

7 *Typical butt joint configurations for sheet and plate material: a) Square edge closed butt <3mm plate thickness; b) Square edge open butt <3mm plate thickness; c) Single V butt, 3–10mm plate thickness; d) Double V butt, >10mm plate thickness.*

backing bar is advisable when welding with an open root joint preparation. The bar normally contains a shallow groove along its length, as shown in the typical design in Fig. 8a, which is suitable for section thicknesses up to 1.5mm (0.06in). Bars may be stainless steel, copper or mild steel. When fabricating components for service in specific environments, choice of material may be restricted to avoid contamination of the weld metal. For example, in nuclear applications where copper contamination must be avoided, bars of a similar composition or ceramic coated steel strip have been successfully employed.

When welding high integrity components, a shielding gas is normally used to protect the underside of the weld pool and the weld bead from atmospheric contamination. Several techniques are available such as gas ports in the backing bar, localised gas shrouds for sheet, or total coverage of the tubular joints using dams or plugs; a commonly used technique for tubular components is shown in Fig. 8b.

When using the plug technique the plugs must be at a sufficient distance from the joint to avoid damage from the heat generated during welding. It is

26

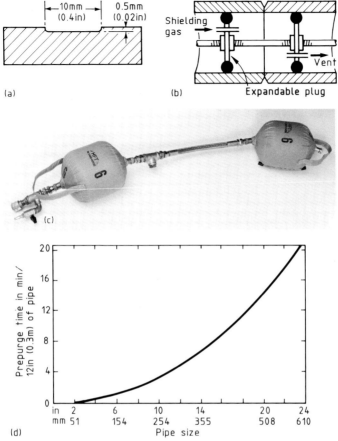

8 *Typical backing systems used in TIG welding: a) Temporary backing bar for sheet material; b) Gas backing for protection of the underside of a tubular weld; c) Pipe purging bladder system; d) Recommended pre-weld purging time (c and d courtesy of Huntingdon Fusion Techniques Ltd).*

also essential with all gas backing systems that the vent hole is sufficiently large to accommodate the gas flow, thus eliminating any tendency to pressurise the weld pool during or on completion of welding. Failure to take account of the pressure build-up results in a concave underbead profile and even expulsion of the weld pool on completion of the weld.

The backing gas used when welding ferritic steels, stainless steels and nickel alloys is normally argon. Nitrogen can be used as a general purpose backing gas for copper. The flow rate depends on the diameter and length of the pipe and is set to ensure five to six volume changes of the pipe system, which should ensure that the oxygen content is reduced to less than 1%. The gas flow rate increases with increased pipe diameter, from typically 10 l/min (20cfh) for a 75mm (3in) diameter pipe to 17 l/min (35cfh) for a 500mm (20in) diameter pipe.

When welding aluminium a backing bar is almost always used in preference to gas backing. For most applications, mild steel is used for the bar, but for high production rates and where weld quality is critical, stainless steel can be used. If the backing is not grooved, it is necessary to back chip to sound metal and to re-weld the root with a sealing run.

Material considerations

The TIG process is more widely applied for welding alloyed steel, nickel alloys, aluminium and the reactive metals titanium and zirconium, and less extensively for mild and carbon steels. Rimming steels, in particular, can suffer from outgassing and porosity unless a filler wire containing deoxidants is used.

Although the choice of shielding gas is largely influenced by the material composition (Table 3) it is worth emphasising that whilst argon is the most common shielding gas, argon-hydrogen and helium-argon mixtures can often increase welding speed. The addition of hydrogen produces a cleaner weld but its use for welding thick section low alloy and high carbon steels is not recommended because of the risk of hydrogen cracking.

Cleaning

Because of the inherent risk of porosity in TIG welding, thorough cleaning of the joint area is essential to remove all traces of oxide, dirt and grease. The normal practice is to use a stainless steel wire brush for carbon, low alloy and stainless steels. A bronze wire brush can be used for copper and its alloys, whilst chemical etch cleaning of aluminium or scraping of the immediate joint area can be particularly effective when welding the reactive metals e.g. aluminium, titanium and zirconium.

It is recommended that the wire brushes used for cleaning the joint preparations should be reserved for the various material types.

Immediately before welding, the weld area should be first degreased with petroleum ether or alcohol and again after wire brushing; degreasing before wire brushing prevents contamination of the wire brush. It is also desirable to scratch brush the joint after each weld pass to remove any oxide film formed during welding.

Finally, it is noteworthy that equal attention should be paid to cleaning the filler wire which must be degreased. Wire baking should not be necessary if stored correctly.

Manual welding

The compactness and lightness of the torch make the TIG process ideal for manual welding of thin components where there is a prime requirement for precise control of the behaviour of the weld pool. High quality welds can be readily achieved even where there is limited access to the joint.

Welding techniques

Significant welder skill is demanded by the TIG process, as the operator must maintain a constant electrode to workpiece distance of approximately 1.6mm (0.06in). Because of the conical arc, torch to workpiece variations as small as 0.5mm (0.02in) can vary the effective area of the arc by as much as 15%. The protrusion of the electrode from the end of the gas shield also gives a risk of electrode contamination, or tungsten inclusions in the weld metal from touching either the weld pool or the filler wire.

When using filler, the rod or wire must be positioned so that its tip is heated by the arc but final melting and transfer of metal occur when the rod is dipped into the weld pool. Thus, to avoid oxygen and nitrogen pick-up, the welder must always ensure that the hot tip of the rod is held within the protective envelope of the gas shield.

The recommended welding techniques i.e. torch and filler rod angles for the various positions for butt welding are shown in Fig. 9 and for T joints in Fig. 10. Welding is preferably carried out in the flat position where gravity has a minimal effect on the behaviour of the weld pool and the highest speeds can be achieved.

In manual welding of pipes the recommended torch and filler rod positions are shown in Fig. 11 for the rotated pipe (1G position) and the fixed pipe (5G position) operations. In this case welding is normally carried out from the six o'clock to the 12 o'clock position.

Special mention must be made of the two-operator technique which is similar to that used in gas welding. It can be employed when welding large components in the vertical position; section thickness must generally be greater than 5mm (0.2in), and the advantages over single operator welding are:

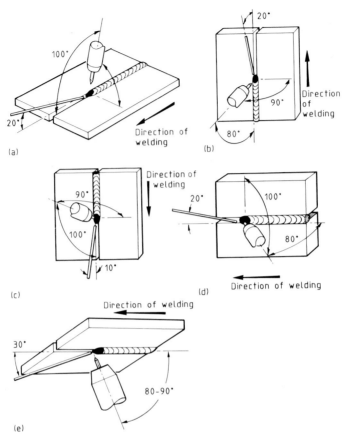

9 Recommended torch and filler rod positions for manual welding butt joints in the following positions: a) Flat; b) Vertical-up; c) Vertical-down; d) Horizontal-vertical; e) Overhead.

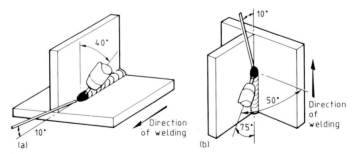

10 Recommended torch and filler rod positions for manual welding T joints in the following positions: a) Horizontal-vertical; b) Vertical-up

30

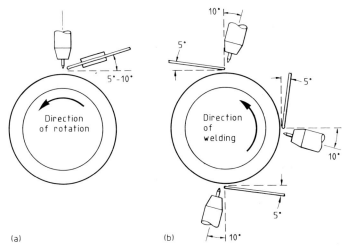

11 *Recommended torch and filler rod positions for manual welding of pipe: a) Pipe rotated (1G) position; b) Fixed pipe (5G) position.*

1 Higher welding speeds;

2 Lower overall welding currents;

3 Smaller weld beads;

4 Reduced joint preparation;

5 Lower distortion.

Welding speeds are approximately twice that of single-operator welding and because the joint preparation is generally smaller filler rod consumption is significantly reduced; a square edge preparation can be used for thicknesses up to 8mm (0.3in) with a double V, 70–80° included angle, for thicker plate material.

Typical welding parameters

Typical welding parameters are given in Table 4a and b for welding butt joints in the flat position in mild steel and stainless steel, respectively. Standard data are also presented on the amount of filler rod consumed, weight of weld metal deposited and arc time per metre of weld.

Applications

Noteworthy examples are welding of aluminium-magnesium alloy piping in chemical plant construction (Fig. 12a) and fabrication of aeroengine

Table 4 Typical welding parameters and standard data for welding butt joints in the flat position

a) Mild steel

Sheet thickness, mm	0.7	0.9	1.0	1.2	1.6	2.0	2.5	3.0
Gap, mm	0.0	0.0	0.0	0.0	0.0	0.0	1.0	1.0
Filler rod diameter, mm	1.6	1.6	1.6	1.6	1.6	1.6	2.4	2.4
Current, A	45	65	70	85	105	125	155	155
Voltage, V	8	9	9	9	9	10	10	10
Filler rod consumed per metre of weld, m, mean	0.79	0.99	1.05	1.20	1.40	1.61	1.54	1.54
, range	1.12	1.33	1.38	1.53	1.74	1.94	1.77	1.77
	0.46	0.66	0.71	0.86	1.07	1.27	1.32	1.32
Weight of weld metal deposited per metre, g, mean	8.2	9.7	10.4	11.8	14.7	17.6	44.0	47.6
, range	14.1	15.6	16.3	17.7	20.6	23.5	49.9	53.5
	2.4	3.8	4.5	5.9	8.8	11.7	38.1	41.7
Arc time per metre of weld, min, mean	3.87	3.33	3.46	3.41	3.75	3.98	4.27	5.27
, range	4.74	3.86	3.97	3.81	4.09	4.28	4.82	5.85
	3.35	3.00	3.15	3.15	3.52	3.78	3.93	4.90

b) Stainless steel

Sheet thickness, mm	0.7	0.9	1.0	1.2	1.6	2.0	2.5	3.0
Gap, mm	0.0	0.0	0.0	0.0	0.0	0.0	0.0	0.0
Filler rod diameter, mm	1.6	1.6	1.6	1.6	2.4	2.4	2.4	2.4
Current, A	55	95	105	135	175	185	195	195
Voltage, V	9	10	10	10	10	11	11	11
Filler rod consumed per metre of weld, m, mean	0.83	1.08	1.14	1.33	1.02	1.06	1.10	1.10
, range	1.12	1.36	1.43	1.61	1.21	1.25	1.29	1.29
	0.55	0.79	0.86	1.04	0.83	0.87	0.91	0.91
Weight of weld metal deposited per metre of weld, g, mean	8.3	9.1	9.5	10.3	29.9	31.5	33.6	35.6
, range	13.4	14.2	14.6	15.5	35.1	36.7	38.8	40.8
	3.1	3.9	4.3	5.1	24.7	26.4	28.4	30.4
Arc time per metre of weld, min, mean	2.91	1.82	1.83	1.62	1.70	2.26	2.92	3.78
, range	3.47	2.07	2.06	1.79	1.83	2.40	3.07	3.95
	2.53	1.65	1.68	1.50	1.61	2.16	2.81	3.64

Note: 1 Shielding gas argon, 2 Joint preparation, square edge

12 *Manual TIG welding applications in a fixed position: a) Aluminium – magnesium alloy piping (courtesy of Air Products Ltd); b) Exhaust diffusion vane to cone manufactured in a nickel based alloy (courtesy of Rolls-Royce Plc).*

13 *Manual TIG welding in which the component or vessel can be rotated: a) Cylinder for a vacuum chamber; joint preparation as shown in Fig. 14c. Welding conditions: shielding gas – argon, backing gas – argon, electrode diameter – 2.4mm, electrode tip angle – 60°, welding current, root – 130A, welding current, fill – 170A, filler rod diameter, root – 1.6mm, filler rod diameter, fill – 2.4mm; b) Aluminium – magnesium pressure vessel using the two-operator technique (courtesy of Air Products Ltd).*

components (Fig. 12b) where in each case the welder has to contend with the problem of difficult access to the joints. Because of precise control over penetration of the weld pool, a skilled welder can produce 'defect free' welds with close control of the weld bead profile.

A typical example of high quality pipe welding for a vacuum chamber is shown in Fig. 13a. Welding was carried out using the simplest technique,

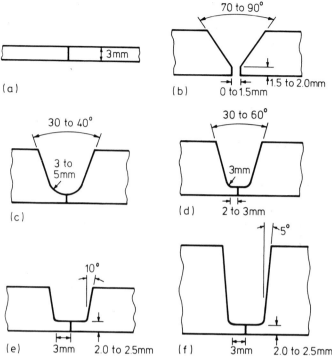

14 *Typical joint preparations used in welding tubes. The joint configurations are for guidance only as the dimensions may vary, and other configurations are shown in succeeding figures to illustrate specific applications: a) Simple butt, <3mm wall thickness, manual or mechanised; b) V type, >3mm wall thickness, manual or mechanised operation; c) U type, >3mm wall thickness, usually manual; d) U type with extended 'land', usually mechanised; e) U type orbital welding, mechanised operation; f) Narrow gap, mechanised operation.*

i.e. by rotating the pipe under a fixed torch position. The joint preparation was of the U type, Fig. 14c, which enabled the underbead profile to be controlled precisely and minimised the filler required to complete the joint.

In an example of two-operator welding of a 6mm (0.25in) wall, aluminium magnesium pressure vessel, Fig. 13b, the vessel was rotated so that welding could be carried out in the vertical position.

The TIG process is often used for the root pass only, with subsequent joint filling carried out using MMA or MIG. By adopting this approach, complete root fusion can be more readily achieved, and the use of a higher deposition rate process for the filler passes ensures that the joint is filled as efficiently as possible.

Automatic wire feed

Special mention should be made of the use of equipment which automatically feeds the filler wire into the weld pool. This is often referred to as semi-automatic welding; typical commercially available equipment is shown in Fig. 15a.

Whilst the technique hinders manipulation of the weld pool to some extent, e.g. using the wire to push the weld pool through in the root pass, it can be used effectively in the filler passes to achieve a continuous welding operation. An application which exploited the advantage of automatic feed to increase the duty cycle was the strengthening of the vacuum vessel, Fig. 15b, of the Joint European Torus is shown in Fig. 15c. The seams required

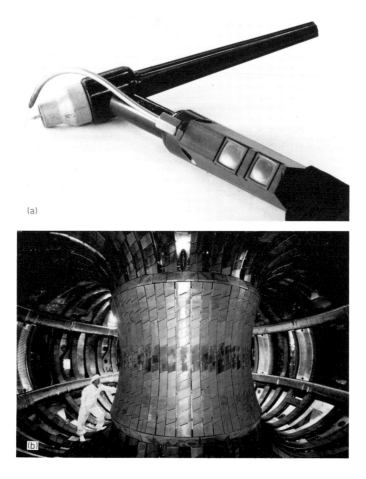

(a)

(b)

15 *Semi-automatic TIG welding: a) Equipment for automatic feeding of the filler wire; b) Vacuum vessel of the Joint European Torus (JET) ; c) Manual TIG welding of the stainless steel inner lining of the toroid vessel (b and c courtesy of the JET Joint Undertaking).*

continuous welding of typically 12mm (0.5in) thick stainless steel and Inconel over 10m (33ft) in length; at a welding current level of 180A, wire feed speeds of over 1000 mm/min (40 in/min) were achieved in the vertical position using a 1.2mm (0.048in) diameter wire.

Mechanised operation

Welding techniques

The TIG process is used extensively in mechanised welding, where high weld quality must be consistently achieved. For instance, in the chemical, aeroengine and power generation industries, the compactness of the torch has been fully exploited in the design and construction of specialised welding equipment. However, because of the inflexibility of mechanised systems compared with manual welding, closer tolerances must be placed on component dimensions and joint fit-up. As a general guide, joint fit-up (gap and vertical mismatch) should be ≯15% of sheet thickness for the normal range of materials and ≯10% for material of less than 1mm. Consequently, when welding thin material it is prudent to devise a good clamping arrangement, preferably using 'finger' clamps to ensure a uniform heat sink along the joint. Whilst joint gap variation is not normally a problem, in butt welding of tubes it is sometimes necessary to size the ends of the tubes to remove excess ovality.

37

In tube welding standard equipment which can accommodate a wide range of tube sizes is currently available for butt welding in all positions. Whenever possible, welding is carried out with the torch in the fixed vertical position, i.e. with the tube rotated beneath the torch (1G welding position) or with the tube positioned vertically (2G welding position). Integral filler, a ring insert or a separate wire feed addition can all be used to enable tolerances on component fit-up to be relaxed. For example, in welding 32mm (1.28in) OD, 3mm (0.12in) wall thickness boiler tubes, whilst the maximum wall thickness variation, without filler wire, was found to be 0.125mm (0.005in), the addition of filler material allowed the variation to be increased to 0.5mm (0.02in).

The simple butt, V and U type joint preparations, (Fig. 14) are all used for welding tube in the flat position; V and U type joints must be adopted for wall thicknesses above 3mm (0.12in) and the U type joint, although more expensive to machine facilitates control of weld bead penetration in the root and reduces the number of weld passes, particularly in thicker wall material. The preferred joint preparation for wall thicknesses above 3mm (0.12in) has a root face of 2mm (0.08in) and a total land width of typically 6mm (0.25in). The wide land is essential to avoid weld pool touching the sidewalls when the pool surface tension could cause suck-back especially in positions other than the flat.

Pulsed current operation

The pulsed current technique has several advantages in mechanised welding especially with regard to improving tolerance to material and production variations. Specific advantages of pulsed operation are as follows:

1 It aids control of weld bead penetration by increasing tolerance to process variations (component dimensions and joint fit-up).

2 It reduces the sensitivity to a disparity in heat sink, for example, in components requiring a weld to be made between thin and thick sections.

3 In certain materials it can reduce the sensitivity to surface oxides and to cast to cast compositional variations.

4 It helps to reduce distortion in thin section material or through poor clamping.

It should be noted, however, that pulsing the welding current inevitably reduces welding speed as the weld pool is allowed to freeze between pulses. For example, with continuous current operation 2.5 mm (0.1in) thick,

16 *Mechanised welding of a heatshield assembly in a nickel based alloy (courtesy of Rolls-Royce Plc).*

stainless steel sheet can be welded at 0.5 m/min (20 in/min) travel speed, but with pulsed current operation, and allowing the weld pool to solidify between pulses, the travel speed would be reduced to 0.12 m/min (5 in/min). Typical pulse parameters for a range of thicknesses of stainless steel are given in Fig. 5.

Applications

Compared with manual welding, mechanised welding is used as a means of increasing welding speed and to achieve more consistent results over long lengths or in mass production. However, if a high level of weld quality is to be maintained, greater attention must be paid to the accuracy of joint fit-up and the welding parameters. A typical example of precision mechanised welding in aero engine manufacture is shown in Fig. 16.

As described in Chapter 1, pulsing is often used in welding thin sheet material, for the root pass of thicker material, and in difficult welding positions, as a means of improving the tolerance of the process. Applications in tube and tube to plate welding are described more fully in subsequent chapters.

A notable example in which pulsing was used to control the weld pool profile when welding in the more difficult 2G position was joining a pendant liner to a fueling standpipe extension and the reheater tube to tubeplate welds for the AGR. The joint configuration for the pendant liner is shown in Fig. 17a. The joint had to be welded on site in the 2G position; an integral filler was provided to ensure that no thinning of the wall occurred. In the initial technique a ceramic covered backing strip was attached to the outside of the joint to protect and support the weld pool during welding. Because of variations in the underbead profile through localised contact

39

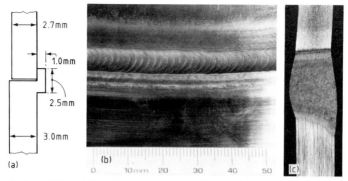

17 *Pulsed TIG welding of pendant liner to fueling standpipe extension using a gas backed technique: a) Joint configuration; b) General appearance of weld; c) Section through weld (courtesy of Darchem Engineering Ltd). Typical welding parameters: shielding gas – argon, pulsed current – 100A, pulsed time – 1sec, background current – 20A.*

with the strip, a two run technique was required to even out weld bead penetration. A conventional gas backing system was found to be preferable, as not only could the weld be completed in a single pass but the underbead surface had a much cleaner appearance. The resultant surface appearance and cross section through the pulsed TIG weld (gas backed) are shown in Fig. 17b and 17c respectively.

Typical defects

The type of defects and their characteristic appearance are listed in Table 5. Information is also presented on the likely cause of the defects and possible remedial actions.

The main problem in TIG welding is maintaining a uniform degree of penetration. Difficulties occur if adequate attention is not paid to minimising variations which may arise in production such as:

Process parameters

Component dimensions
Joint fit-up

Welding parameters

Welding current
Electrode – workpiece distance
Electrode dimensions
Shielding gas composition
Shielding gas flow rate
Welding speed

Table 5 Typical defects

Defect	Appearance	Cause	Remedy
Lack of root penetration	Notch or gap	Current level too low	Increase current
		Welding speed too high	Decrease welding speed
		Incorrect joint preparation	Increase joint angle or reduce root face
		Arc too long	Reduce arc length
	Concave underbead	Tacks not fully fused	Reduce size of tacks
		In flat position, backing gas flow too high	Reduce backing gas flow rate
		In position, intolerant joint preparation	Use U preparation and ensure weld pool does not bridge the sidewalls
Lack of side-wall fusion	Not normally visible, detected by NDT (radiography and ultrasonic examination) or side-bend tests	Current level too low	Increase current level
		Welding speed too high	Decrease welding speed
		Incorrect torch angle	Incline torch backwards and hold arc on leading edge of weld pool
		Incorrect joint preparation	Increase joint angle
		Too large rod/wire diameter for plate thickness	Reduce rod/wire diameter
		Insufficient cleaning	Clean plate surface
Undercut	Groove or channel along one edge of weld	Welding current too high	Reduce welding current
		Welding speed too high	Reduce welding speed
		Torch inclined to one side	Incline 90° to plate surface
Porosity	Severe case, surface pores but normally sub-surface detected by radiography	Insufficient shielding	Increase flow rate (see Fig. 3 for correct flow rates)
		Turbulence in the shield	Decrease flow rate
		Disturbance of the shield through draughts	Shield joint area
		Dirty plate material, e.g. oil, grease, paint	Clean surfaces and remove degreasing agent
		Dirty wire material	Clean wire and remove degreasing agent
		Contaminated gas	Change gas cylinder. Purge gas lines before welding. Check connections. Use copper or Neoprene tubing
Weld metal cracks	Crack along centre of weld	Excessive transverse strains in restrained welds	Modify welding procedures to reduce thermally induced strains
		Low depth to width ratio (D:W)	Adjust parameters to give D:W of 1:1
		In autogenous welding, incorrect parent metal composition	Reduce sulphur, phosphorus contents to <0.06% total
		Surface contaminants	Clean surfaces, particularly remove cutting lubricants
		Large gaps in fillets welds	Improve joint fit-up

41

Material

Cast composition
Surface cleanness

In manual welding, weld quality is very much in the hands of the welder and skill is required to ensure that full penetration is achieved without lack of fusion or porosity defects in the body of the weld.

In mechanised welding, where the operator has little control of the behaviour of the weld pool, greater control must be exercised over the component dimensions and fit-up. It is also essential that stringent control over the equipment settings and welding parameters is assured by means of

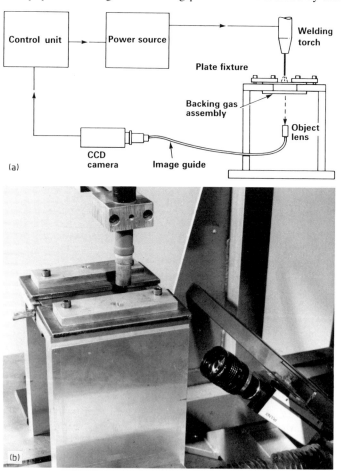

(a)

(b)

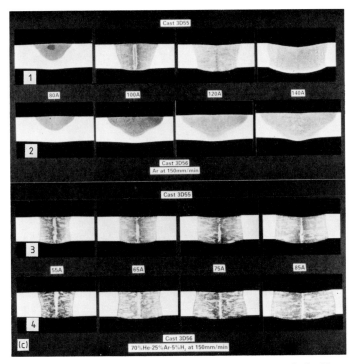

18 *Control of penetration in TIG welding: a) System arrangement for backface control; b) Direct observation of backface of weld using CCD camera; c) Sections through 'good' and 'poor' casts of stainless steel showing the beneficial effect of helium based shielding gas on weld bead penetration profile: 1 – 'Good' cast argon shielding gas, showing effects of current level on depth of penetration; 2 – 'Poor' cast, argon shielding gas, depth of penetration for the same range of welding currents; 3 – 70% helium – 25% argon – 5%H_2 shielding gas showing penetration behaviour for the 'good' cast; 4 – 70% helium – 25% argon – 5%H_2 shielding gas showing 'poor' cast to have almost identical penetration behaviour as 'good' cast.*

a quality control scheme. Instrumentation packages are now available commercially for monitoring the performance of welding equipment during production. Weld pool penetration control systems are also available commercially which are capable of ensuring that the weld pool completely penetrates the material. A novel technique is the use of a CCD camera focused on to the back of the weld as shown in Fig. 18 a and b. The control system chops the peak current pulse in response to a preset quantity of light emanating from the weld pool. The system has applications in the manufacture of critical components for the nuclear and aero engine industries where full fusion must be guaranteed.

Material variation

Special mention must be made of variations in penetration which are caused by minor differences in material composition. Generally known as cast to cast variation, two heats of material conforming to the same nominal specification may produce vastly differing weld bead shapes when welded with exactly the same welding procedure. The problem has been attributed to small differences in the level of impurity elements in the material. For example, casts of austenitic stainless steel which are low in sulphur, typically less than 0.008% tend to display poor penetration behaviour; transverse sections through orbital TIG welds of good (0.014%S) and poor (0.002%S) casts of stainless steel are shown in Fig. 18c, 1 and 2, respectively.

Two process techniques which are capable of improving the tolerance of the TIG operation to variations in material composition are low frequency current pulsing (see Chapter 1) and the selection of the shielding gas composition. With regard to the latter, helium based mixtures have been found to be very effective, especially a three component shielding gas, 70% helium/25% argon 5% hydrogen. As shown in Fig. 18c, 3 and 4, the penetration behaviour of the above two casts now performed almost identically when welded using the helium based shielding gas.

However, it must be emphasised that helium rich gases are not always successful in overcoming the effect of material variation, and it is advisable to investigate other shielding gas mixtures, for example argon-H_2 mixtures, and other combinations of welding parameters especially lower welding speeds.

Further reading

1 Lucas W and Rodwell M R: 'Improving penetration control in TIG welding' *Welding Review* 1987 6 3.

2 Heiple C R and Roper J R: 'Mechanism for minor element effect on GTA fusion zone geometry' *Weld J* 1982 61 4.

3 Hall E T 'The evolution and application of TIG-welding in the manufacture of fabrications for gas turbine aeroengines'. TIG and plasma welding, The Welding Institute, 1978.

4 Bromwich R A C: 'Automatic TIG-welding in fabrication and repair of power plant'.

5 Anderton J G: 'Some TIG-welding applications in the aerospace industry'.

6 Normando N J: 'Manual pulsed GTA welding' *Weld J* 1973 52 9.

7 Glickstein S S and Yeniscavich: 'A review of minor element effects in the welding arc and weld penetration'. Res Council Bull No. 226 May, 1977.

8   Pollard B: 'The effects of minor elements on the welding characteristics of stainless steel'. *Weld J* 1988 67 9.

9   French W: 'Small bore tube processing lines' *Metal Construction* 1984 16 7.

10 Nomura H, Fujiaka T, Wakamatsu M and Saito K: 'Automatic welding of the corrugated membrane of an LNG tank'. *Metal Construction* 1982 14 7.

11 Woolcock A and Ruck R J 'Argon shielding techniques for TIG welding titanium' *Metal Construction* 1980 12 5.

12 Sewell R A 'Gas purging for pipe welding'. *Welding and Metal Fabrication* 1989 57 1.

13 Hall E T 'The evolution and application of TIG welding in the manufacture of gas turbine aeroengines'. TIG and plasma welding, The Welding Institute, 1978.

14 Wareing A J: 'Control of TIG welding in practice', *Welding and Metal Fabrication* 1988 56 8.

15 Wright K: 'Application of automatic tube welding in boiler fabrication at FWPP Ltd'. Advances in Process Pipe and Tube Welding, Abington Publishing.

CHAPTER 3
Mechanised orbital tube welding

Welding techniques

When it is not possible to rotate the tube, as in the construction of chemical plant or boilers, orbital welding techniques can be applied.

Although a simple V preparation of 70–90° included angle and 0.5 (0.02in) to 1.0mm (0.04in) root face (Fig. 14b) can be used, a pre-placed insert can be employed to improve the uniformity of the root penetration. A number of inserts are shown in Fig. 19a, and whilst some of the designs are self-locating, it is prudent to match the tube ends and to tack the insert in position before welding. The appearance of the more common EB insert pre-tacked into the joint is shown in Fig. 19b. Despite the use of a simple V

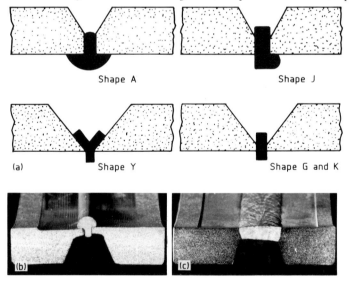

Shape A Shape J

(a) Shape Y Shape G and K

(b) (c)

19 *The use of a consumable insert in butt welding tubes to improve the uniformity of root penetration: a) Insert designs; b) General appearance of Shape A, often referred to as an EB insert, tacked into place; c) Appearance of penetration bead in the overhead position.*

preparation, the additional material provided by the insert was sufficient to avoid suck-back of the root pass in the overhead position, Fig. 19c.

However, if the welding parameters are not carefully set, typical root defects which can still be experienced, include:

- Incomplete root fusion;
- Uneven root penetration profile;
- Root penetration concavity (suck-back).

The most tolerant joint preparation for wall thicknesses above 3mm(0.12in) is the U-shaped preparation (Fig. 14e) with a root face of 2.0 (0.08in) to 2.5mm (0.1in) and a total land width of 6mm (0.25in) between the sidewalls. This preparation not only reduces the filling required (in thicker wall tube) but also assists in containing the weld pool in the vertical position. However, because of the closeness of the sidewalls, the torch must be accurately tracked along the joint; poor tracking may result in suck-back in the root pass or lack of sidewall fusion defects in the filler passes.

Pulsing the welding current is particularly useful in ensuring that a positive penetration bead is obtained in all welding positions. However, when suck-back problems are experienced additional process techniques can be adopted. For example, pulsing the wire feed in synchronism with the background current period has been found to be especially effective for the following reasons:

- When no filler wire is added during the high current period the arc current can be held at a level sufficient to give penetration;
- Feeding of wire during the background period rapidly freezes the weld pool;
- Weld penetration bead is pushed through and supported whilst it freezes.

The disadvantage of this technique is that, as the welding procedure is more difficult to set up, greater operator training is required.

Another process technique which has been successfully applied in production is negative purging. In this technique the backing gas is reduced to slightly less than atmospheric pressure, so that in effect the weld pool is drawn into the bore of the tube.

Equipment

Standard commercial welding systems are currently available for butt welding in all positions, for a wide range of tube sizes, and the associated

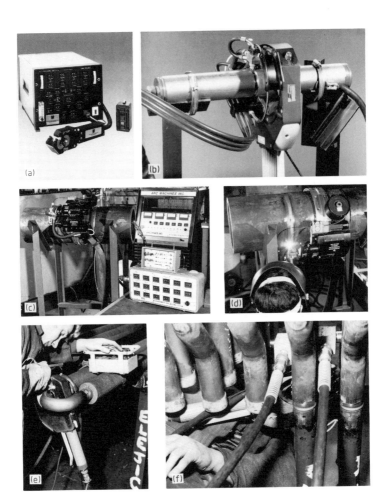

20 *Typical commercially available orbital tube welding equipment: a) General arrangement of basic function system (courtesy of Huntingdon Fusion Techniques Ltd); b) Basic function welding head (courtesy of ESAB Ltd); c) Full function welding system; d) Full function welding head (courtesy of Arc Machines Inc); e) Production welding of boiler tubes; f) Production welding with restricted access (courtesy of Foster Wheeler Power Products Ltd).*

power sources have the capacity to programme the welding parameters for various welding positions (vertical, overhead, etc) around the joint and to pulse the welding current; examples of commercially available systems are shown in Fig. 20.

The welding equipments vary quite considerably both in terms of the operating features and indeed the price. The simple systems have been termed basic function and are characterised by the restriction to stringer bead welding i.e. no electrode oscillation, although current pulsing is usually provided (Fig. 20a). The full function systems contain the following features:

1 Electrode oscillation;

2 Pulsed current synchronised to electrode oscillation;

3 Pulsed wire feed;

4 Pulsed travel;

5 Multi-level programming;

6 Automatic arc voltage control (AVC).

Intermediate systems contain several but not all of these features. Examples of typical welding heads for the basic function and the full function systems are shown in Fig. 20b, c and d respectively.

The most important limitation of the basic function systems is that the operator is restricted to a stringer bead welding technique, in comparison to the intermediate and full function systems. This necessitates more welding passes, reduction in the tolerance to variation in joint dimensions and fit-up and the increased weld contraction may cause excessive bore constriction.

In addition to the benefits of electrode oscillation, the full function systems, in particular, allow the following beneficial welding techniques in control of the weld pool:

- In the root pass, pulsed wire feed with the wire added during the background period promotes rapid freezing and control of the weld pool during this part of the cycle.

- In the hot pass, synchronised pulsing of the current so that a high current level is applied during the end-dwell period improves sidewall fusion; the synchronisation of the low current period on the centre of the joint reduces the risk of re-penetration through the root pass.

- When filling the joint the use of arc voltage control is especially useful as the arc length can be maintained despite the pipe ovality and variations in the weld bead contour.

Table 6 Typical joint preparation and welding parameters for welding 50mm (2in) OD, 4mm (0.16in) wall thickness, carbon steel pipe in the flat (5G) position (courtesy of Foster Wheeler Power Products Ltd)

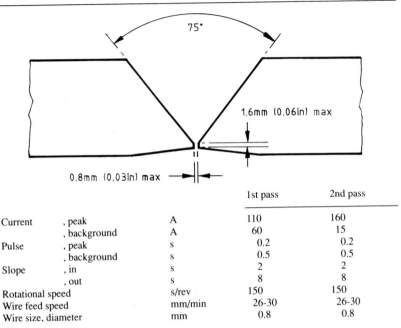

			1st pass	2nd pass
Current	, peak	A	110	160
	, background	A	60	15
Pulse	, peak	s	0.2	0.2
	, background	s	0.5	0.5
Slope	, in	s	2	2
	, out	s	8	8
Rotational speed		s/rev	150	150
Wire feed speed		mm/min	26-30	26-30
Wire size, diameter		mm	0.8	0.8

The special techniques afforded by these additional control features facilitate the production of sound welds but it should be noted that in addition to the higher price, the heads are larger and greater operator training is required.

Special purpose equipment is also available for example, for welding tubes with as little as 50mm (2in) clearance between them, and despite the restricted access automatic arc length control and wire feed features have been included in specific designs.

Applications

Mechanised orbital welding techniques are used to achieve a more consistent level of weld quality, the tube edge preparation should always be machined so that close tolerances are achieved. Examples of the use of basic function orbital TIG welding in the manufacture of boilers are shown in Fig. 20e and Fig. 20f with the latter being an application with very restricted access; typical joint preparation and welding parameters are given in Table 6. Welding of typically 50mm (2in) OD x 4mm (0.16in) weld thickness, carbon steel pipe was carried out using a basic function system and the weld was completed in two passes.

Electrode oscillation has been used when welding 60mm (2.5in) OD x 5.7mm (0.23in) wall stainless steel pipe to reduce the number of passes from 8 to 4, in Fig. 21. Furthermore, using the stringer bead procedure (basic function system) the bore constriction of typically 1.9mm (0.076in) was outside the maximum of 1.5mm (0.06in) permitted by BS 4677: 1984.

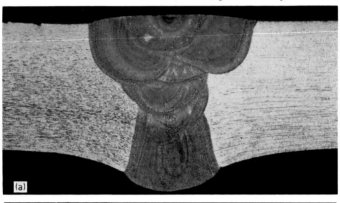

Welding parameter	Unit	Pass number							
		1	2	3	4	5	6	7	8
Wire diameter	mm	0.8	0.8	0.8	0.8	0.8	0.8	0.8	
Wire feed rate	m/min	0.26	0.26	0.31	0.31	0.25	0.21	0.13	
Pulsed peak current	A	72	76	98	100	100	100	100	70
Pulse time	sec	0.8	0.8	0.8	0.8	0.8	0.8	0.8	0.8
Background current	A	29	30	40	40	40	40	40	28
Background time	sec	0.4	0.4	0.4	0.4	0.4	0.4	0.4	0.4
Motor delay	sec	5	5	5	5	5	5	5	5
Slope up	sec	1	1	1	1	1	1	1	1
Slope down	sec	6	6	6	6	6	6	6	6
Rotation speed	sec/rev	160	160	160	160	160	160	160	160
Shielding gas flow rate	l/min	7	7	7	7	7	7	7	7
Purge gas flow rate	l/min	3	3	3	3	3	3	3	3

Electrode type	2% thoria
Electrode diameter	2.4 mm
Electrode angle	60°
Electrode polarity	DC−
Shielding gas	Ar-1%H$_2$
Purge gas	Argon

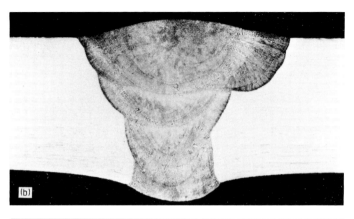

		Pass number			
Welding parameter	Unit	1	2	3	4
Wire diameter	mm	0.8	0.8	0.8	0.8
Wire feed rate	m/min	0.75	0.75	0.90	0.75
Welding current	A	81	95	105	80
Weave amplitude	mm	2.0	2.5	3.0	5.0
Weave rate	mm/sec	10	10	10	10
Weave frequency	Hz	0.36	0.38	0.48	0.52
End dwell	sec	0.4	0.3	0.3	0.3
Motor delay	sec	6	6	6	6
Slope up	sec	1	1	1	1
Slope down	sec	6	6	6	6
Rotation speed	sec/rev	165	165	165	200
Shielding gas flow rate	l/min	7	7	7	7
Purge gas flow rate	l/min	3	3	3	3

Electrode type	2% thoria
Electrode diameter	2.4 mm
Electrode angle	60°
Electrode polarity	DC–
Shielding gas	Argon
Purge gas	Argon

21 *Sections through tube welds in 60mm OD and 5.7mm wall (2in NB Schedule 80), type 304 stainless steel pipe in the 5G position: a) Weld with basic function head; b) Weld with electrode oscillation.*

Sections through the welds and the major welding parameters are given in Fig. 21a and b for the stringer and weaved welds respectively.

Because of the reduction in the number of passes there are also significant cost benefits to be gained from using electrode oscillation. A detailed cost analysis using The Welding Institute's WELDVOL and WELDCOST microcomputer program is shown in Table 7. The stringer bead technique requires almost twice as long as the electrode oscillation technique principally because of the extra number of filler passes, and the greater indirect time used to recoil the welding bead cables, modify the welding parameters, clean off the surface oxide between passes and regrind the electrode.

Table 7 Comparison of cost per metre, consumables used and weld times for welding 60mm OD × 5.7mm wall (2in NB, Schedule 80) type 304 stainless steel, using stringer bead and electrode oscillation procedures. (Data produced using The Welding Institute's WELDVOL and WELDCOST programs.)

Stringer bead
Welding cost per metre

Gas	1.0%	1.21
Wire	1.9%	2.27
Electrode		0.00
Rods		0.00
Flux		0.00
Labour	85.1%	100.00
Plant	11.9%	13.95
Power	0.1%	0.08
TOTAL welding cost per metre (pounds)		117.52

Consumables used

Shielding gas used/metre (litre)	= 540.0
Cost (pound/metre)	= 1.08
Backing gas used/metre (litre)	= 66.7
Cost (pound/metre)	= 0.13
Mass of wire used/metre (kg)	= 0.23
Cost (pound/metre)	= 2.27

Weld times

Arc time (min/metre)	= 90.00
Indirect time (min/metre)	= 210.00
Total time (min/metre)	= 300.00

Electrode oscillation
Welding cost per metre

Gas	1.0%	0.92
Wire	2.4%	2.27
Electrode		0.00
Rods		0.00
Flux		0.00
Labour	72.9%	69.44
Plant	23.7%	22.61
Power	0.1%	0.05
TOTAL welding cost per metre (pounds)		95.30

Table 7 *continued*

Consumables used

Shielding gas used/metre (litre)	=	375.0
Cost (pound/metre)	=	0.75
Backing gas used/metre (litre)	=	83.3
Cost (pound/metre)	=	0.17
Mass of wire used/metre (kg)	=	0.23
Cost (pound/metre)	=	2.27

Weld times

Arc time (min/metre)	=	62.50
Indirect time (min/metre)	=	145.83
Total time (min/metre)	=	208.33

Table 8 Parameters for orbital TIG welding of 21.4 mm diameter 70/30 Cu-Ni tube

Constants

Filler composition	70/30 Cu-Ni
Filler diameter	1.2 mm
Shielding gas	argon/5%H_2
Backing gas	argon
Electrode/workpiece gap	2.4 mm
Traverse mode	steady speed
Electrode diameter	1.2 mm
Electrode tip	30° cone

		Filling pass			
	Root	1	2	3	4
Start position, o'clock	11	12	11	12	11
Slope up, A/sec	5.0	20	20	20	10
Pulsed current, A	90	199	199	199	60
Pulse time, sec	0.7	0.2	0.2	0.2	DC
Background current, A	30	30	30	30	DC
Background time, sec	2.0	1.0	1.0	1.0	DC
Filler addition rate, mm/min	140	225	225	225	225
Weave amplitude, mm*	None	None	None	None	5.6
Weave frequency, cycles/min	N/A	N/A	N/A	N/A	50
Weave delay at extremities, sec	N/A	N/A	N/A	N/A	None
Current decay, A/sec	2.5	9.99	9.99	9.99	2.5
Cutoff at, A	30	60	60	60	30
Sequence terminates, A/sec	5.0	12	12	12	5.0
Final current, A	5.0	5.0	5.0	5.0	5.0
Time per revolution, min†	1.1	1.1	1.1	1.1	1.1

*   Weave amplitude measured at electrode tip
†   The rotational speed was kept constant, and therefore the actual welding rate increases as the weld preparation is filled
N/A   Not applicable

54

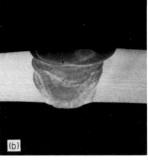

22 *Orbital TIG (GTA) weld in 70/30 Cu-Ni tube; 21.4mm OD, 5.0mm wall thickness: a) Root and completed weld; b) Section through weld. Welding parameters are given in Table 8.*

Another example of orbital welding with the U-shape joint preparation is shown in Fig. 22 where 21.4mm (0.86in) OD, 5.0mm (0.2in) wall thickness 70/30 CuNi tube was welded in the 5G position (welding conditions and parameters are detailed in Table 8). Filler wire was used in the root pass to prevent porosity, and current pulsing was employed throughout, at a frequency of approximately 0.5Hz, to ensure good sidewall fusion. While the pulse current was required to spread the weld pool and to enable the arc to 'bite' into the sidewall, a background period up to 2sec was used to allow the large weld partially to solidify to a more controllable size. Weaving at a frequency of 50 cycles/min was used in the capping pass to spread the weld pool and to prevent undercutting along the edges of the weld bead.

Further reading

1   Stalker W L, Tate E F and Murphy M C: 'Orbital narrow gap welding'. *Metal Construction* 1979 11 4.

2   Dick N T: 'Tube welding by the pulsed TIG method'. *Metal Construction and Brit Weld J* 1973 5 3.

3   Sewell R A: 'Orbital pipeline welding techniques' *Welding and Metal Fabrication* 1989 57 9.

4   Carrick L, Hick A B, Salmon S and Wareing A J: 'A new welding technique for stainless steel pipe butt welds'. *Metal Construction* 1985 17 6.

5   Scott-Lyons R and Middleton T B: 'Orbital TIG system simplifies underwater welding'. *Metal Construction* 1985 17 8 also 'Underwater orbital TIG welding' *Metal Construction* 1984 16 10.

6   Blake M A W, Carrick C and Paton A: 'Automatic TIG welding in site fabrication'. *Metal Construction* 1983 15 5.

7   WELDVOL and WELDCOST - microcomputer packages available from The Welding Institute, UK.

CHAPTER 4
Tube to tubeplate

Welding techniques

The TIG process is widely applied to welding tubes into a tubeplate. Possible joint configurations are shown in Fig. 23. In comparison with other TIG operations, good quality tube to tubeplate welding, with either a front or back face technique, requires greater attention to the initial cleaning of the component surfaces, joint fit-up and to the condition of the electrode tip. With regard to cleaning, low porosity can be achieved by using ultrasonic techniques with final degreasing in a solvent immediately before insertion of the tube into the tubeplate. The fit-up can affect the consistency of penetration, and here roller or hydraulic expansion of the tube into the tubeplate has been shown to be particularly beneficial when using the front face welding technique.

When design considerations permit a front face technique should be employed because:

1 Fewer restrictions are imposed on the design of torch;

2 The electrode is more readily positioned on the joint line;

3 Welding can be directly observed by the welder;

4 Filler can be added easily to the weld pool.

Equipment

Equipment for tube to tubeplate welding usually consists of special purpose machines. The machine for front face welding usually comprises a rotating torch with a means of locating on the edge of the tube. A typical production machine is shown in Fig. 24a. In this case the welding head is located on the tube to be welded by means of a centre mandrel and then fixed in position by means of pneumatically operated 'pull-in' cylinders; the equipment in production use is shown in Fig. 24b.

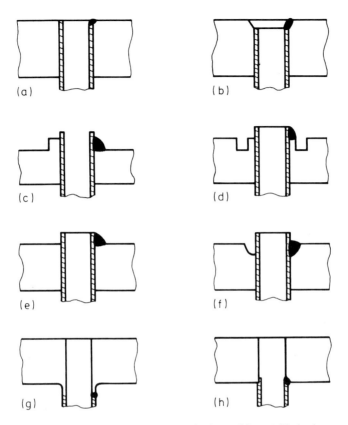

23 *Typical joint configurations for tube to tubeplate welding: a) Flush tube; b) Recessed tube; c) Added ring; d) Trepanned tubeplate; e) Fillet; f) Extended tube; g) Tube to boss; h) Recessed (backface) tube.*

Back face machines are normally more complex as they often incorporate sensor systems for pre-setting the electrode to joint distance; a bore welding torch which was used for welding the steam generator for the advanced gas cooled reactor is shown in Fig. 24c.

Applications

Front face welding

A typical example of front face welding is shown in Fig. 25, in which 13mm (0.5in) OD 2.5mm (0.1in) wall thickness carbon-manganese tube was welded into the tubeplate using the flush tube joint configuration (Fig. 23a). Welding was carried out using argon-helium shielding and a two pass

24 *Examples of tube to tubeplate welding heads: a) Front face torch; b) Front face torch in production use; c) Back face torch (courtesy of Babcock Energy Ltd).*

25 *Front face welded in 13mm OD carbon-manganese tube and tubeplate, joint configuration as shown in Fig. 22a: a) General view; b) Finished weld. Welding conditions: shielding gas – 60% helium/40% argon, electrode diameter – 2.4mm, electrode tip angle – 40°, welding current – 110A, welding speed – 30 sec/rev.*

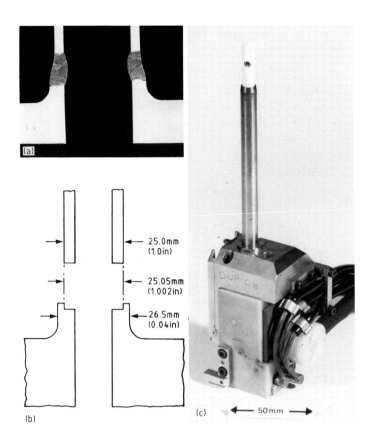

26 *Back face weld in small diameter tube, 25mm (1in) OD x 2.4mm (0.1in) wall thickness, type 347 stainless steel tube to type 347 stainless steel tubeplate: a) Section through weld; b) Joint configuration; c) Welding torch (courtesy of Foster Wheeler Power Products Ltd). Welding conditions: welding speed – 60 sec/rev, welding current – 50A, voltage – 13V, delay – 5sec, rundown – 5sec, arc gap – 12mm (0.5in), electrode position below joint line – 0.04mm (0.002in), welding position – tube vertical.*

technique to avoid weld sinkage, which would have resulted in unacceptable bore protrusion.

Back face welding

Where the major requirement is for a crevice free joint, the back face technique must be adopted using one of the joint configurations shown in Fig. 23g and h. The tube to boss type is more expensive to prepare, and

59

wasteful of material, but it facilitates butt welding and the joint is more readily inspected non-destructively.

An example of a small diameter bore weld in stainless steel is shown in Fig. 26a which required a small precision torch (Fig. 26c) capable of operating within a bore of approximately 25mm (1in.) The joint and welding parameters (given in the figure) were designed to control the degree of bore protrusion and to obtain a smooth external profile with no thinning of the

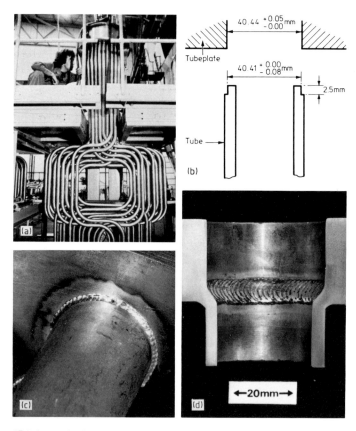

27 *Tube to tubeplate weld for the reheater pod in the Advanced Gas Cooled Reactor: a) General arrangement of the tubes and tube plate; b) Joint configuration; c) Appearance of back of weld; d) Section through weld (courtesy of Babcock Energy Ltd). Typical welding parameters: shielding gas – helium/argon/5%H_2, pulsed welding current – 120A, pulsed time – 2.6sec, background current – 25A.*

tube wall. The outer wall is protected from oxidation by a layer of submerged-arc flux or argon shielding gas.

The reheater feed water pods for the advanced gas cooled reactor are a notable example of TIG welding of tube to tubeplate joints. The general arrangement of a pod is shown in Fig. 27a and the joint configuration in Fig. 27b. Welding was carried out using a sophisticated bore welding torch (Fig. 24c) which, despite the relatively small bore of the tubes (approximately 40mm (1.6in) OD), had facilities for remote setting of the electrode position relative to the end of the tube.

The essential quality requirements of these joints were zero porosity on radiography, no undercutting of the tubeplate wall and no bore protrusion. Inconsistent results were obtained with the continuous current TIG technique, with the weld pool often failing to penetrate or the incidence of excessive penetration and undercutting. The observed variations in the weld bead penetration profile were caused, at least in part, by the inability of the TIG process to accommodate the variations in heat sink (from variations in ligament thickness), arc position and arc length (through distortion of the shape of the tube hole) which inevitably occur in welding this type of joint configuration.

The pulsed mode of operation, however, largely overcomes these difficulties, particularly if the rotation of the torch was also pulsed. Movement of the torch was carried out during the pulse period, i.e. when the arc forces were still effective, so as to minimise the risk of the molten weld pool flooding back on to the electrode. Uniform weld bead penetration was consistently achieved as shown in the general appearance of the back of the weld, Fig. 27c, and the section through the weld, Fig. 27d.

Further reading

1   Schwartzbert H 'In-bore gas tungsten arc welding of steam generator tube to tubesheet joints'. *Weld J* 1981 60 3.

2   Moorhead A J and Reed, R W 'Internal bore welding of $2^{1/}4$ Cr-1Mo steel tube to tubesheet joints'. *Weld J* 1980 59 1.

CHAPTER 5
Micro-TIG welding

Welding techniques

The advent of transistor controlled power sources has facilitated the design of very low current power sources; welding currents of less than 1A can now be reliably initiated and held to an accuracy of ±0.5%. To stabilise the arc it is also necessary to use small diameter 2 or 4% thoriated electrodes and the recommended minimum electrode size for various current ranges are given in Table 9.

A small tip of 8–10° is normally used which facilitates arc initiation and arc stability at the low current levels.

To achieve uniform fusion in thin sheet material, i.e. without burn-through, it is essential that the component edges are accurately machined and that the clamping provides a uniform heat sink. It is also necessary to ensure that the two faces are in intimate contact along the entire length of the joint. Tolerance to joint fit-up can be improved by overlapping the sheets or by preforming the edges as shown in Fig. 28.

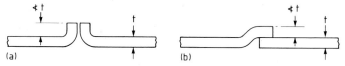

28 *Edge preparation for welding thin sheet materials: a) Flanged edge; b) Micro-lap.*

Table 9 Recommended electrode diameter and vertex angle for micro-TIG welding at various current levels

| Electrode diameter | | Current range |
mm	in	A
0.25	0.010	0–2
0.5	0.020	3–8
1.0	0.040	8–20

A tip angle of 8–10° is normally recommended

Equipment

A suggested clamping arrangement for butt joints in thin sheet is shown in Fig. 29. The recommended jig design and setting dimensions are given in Table 10. The clamps and backing bars should be made of copper or steel with copper toes or inserts. Clamping should be even over the entire seam length and 'finger' clamps are often used to ensure this.

Specialised mechanised equipment is available with such features as automatic electrode positioning, as shown in Fig. 30a.

Applications

Micro-TIG is now replacing micro-plasma for welding thin section components such as diaphragms and bellows, for single-shot spot welding of wires on to pins and for rounding off surgical catheter guide wire ends. A typical example of a micro-TIG welded application is shown in Fig. 30b

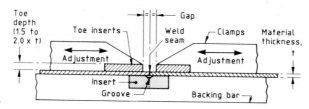

29 *Clamping arrangement for welding thin sheet material (courtesy of Precision Systems Ltd).*

Table 10 Recommended clamping arrangement for micro-TIG welding thin sheets (courtesy of Precision Systems Ltd)

Material thickness		Clamp spacing			
		Min		Max	
mm	in	mm	in	mm	in
0.075–0.5	0.003–0.020	0.75	0.030	2.0	0.080
0.5–2.0	0.020–0.080	2.0	0.080	4.0	0.160
>2.0	>0.080		$1.5–2.0 \times t$		

t – thickness of sheet
Toe depth – 1.5 to 2.0 t
Groove width – $2 \times t$
Groove depth – $1 \times t$ (or 0.025 mm (0.010 in) whichever is the mean)

30 *Micro-TIG welding of a load cell: a) Welding equipment showing power source, workpiece handling and electrode positioning; b) Location of joint between the bellows and the end sections (courtesy of Huntingdon Fusion Techniques Ltd).*

which is a load cell made up of a bellows (50mm (2in) OD, 0.08mm (0.003in) wall thickness) wall stainless steel section welded to an Armco body.

Further reading

1   P W Muncaster: 'Developments in low current DC TIG precision welding'. 'Advanced Welding Systems', int conf, London, Nov 1985, publ The Welding Institute.

2   Donath V: 'How can TIG microwelding be mechanised?'. *Schweisstechnik Berlin* 1977 27 2.

CHAPTER 6

Hot wire TIG welding

Welding techniques

The hot wire TIG variant was developed as a means of achieving very high deposition rates without reducing the high weld quality normally associated with TIG welding.

The essential feature is that filler wire is fed directly into the back of the weld pool and resistance heated using a separate power source which may be AC or DC as shown in Fig.31. An AC power source minimises any interference with the welding arc through the magnetic field generated by the current flowing in the wire and is normally chosen for mechanised systems. However, a single power source is also available commercially for supplying the arc and wire heating currents by a switching arrangement (Fig. 31c). Thus, in operation, the arc melts the base metal to form the weld pool. The filler wire, heated to its melting point source, enters the weld pool behind the arc to form the weld bead, Fig. 32. Smooth feeding of the wire, control of angle of entry into the weld pool, and a stable power source are all essential for stable operation, otherwise random arcing from the filler wire occurs with the resulting pool disturbances causing porosity.

Equipment

Examples of systems commercially available for manual welding and mechanised welding are shown in Fig. 33. In the manual system, a single power source is used (Fig. 33a) and the wire guide tube is attached to the side of the gas nozzle (Fig. 33b). The mechanised system has separate power sources and controllers for the arc and the wire (Fig. 33c) and the gas nozzle for preventing oxidation of the wire is separated from the TIG torch (Fig. 33d).

Applications

The main advantage of the process is that deposition rates can be achieved similar to those obtainable with MIG welding; typical rates are shown in

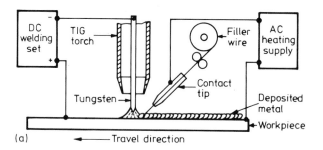

(a) ← Travel direction

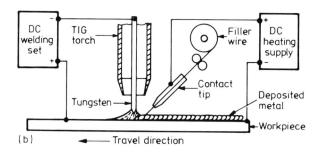

(b) ← Travel direction

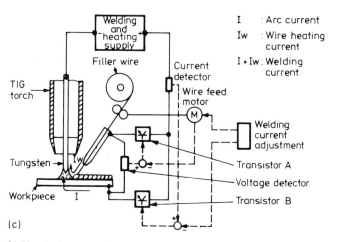

(c)

I : Arc current
Iw : Wire heating current
I + Iw : Welding current

31 *Electrical system and torch arrangement for the hot wire TIG process: a) Separate hot wire AC power source; b) Separate DC power supply; c) Single power source for arc and wire currents.*

32 *Torch – filler wire arrangement showing angle of entry of the wire into the back of the weld pool.*

33 *Commercially available hot wire TIG equipment: a) Manual system with single power source and integral wire feed; b) Manual torch; c) Mechanised system with separate power sources for arc and wire; d) Mechanised head with separate torch and wire feed nozzle.*

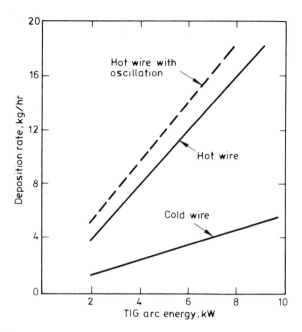

34 *Deposition rates that can be achieved with the hot wire TIG process compared with the conventional TIG cold wire process.*

Fig. 34. As shown in Fig. 35, the process can be used for a range of thicknesses to increase the welding speed whilst retaining good weld penetration and profile characteristics. The process has benefits over MIG for welding stainless steel which normally demands low defect levels and for 9% nickel steels which can suffer from magnetic arc blow.

The process has been used to reduce substantially the number of passes; a typical example is welding cast reformer nickel based alloy tubes, Fig. 36. A typical joint preparation for a weld in 8mm wall thickness, which was welded in three passes, is shown in Fig. 36b. In comparison, welding tube of the same wall thickness with the conventional cold wire TIG technique would have required at least five passes.

In large fabrications, the process has also been used for welding in the vertical-up position using DC current through the wire and mounting the welding head on to a portable carriage and track, Fig. 37. The advantageous features compared with the alternative of TIG-cold wire, are:

1 Deposition rate is increased by a factor of 2.

68

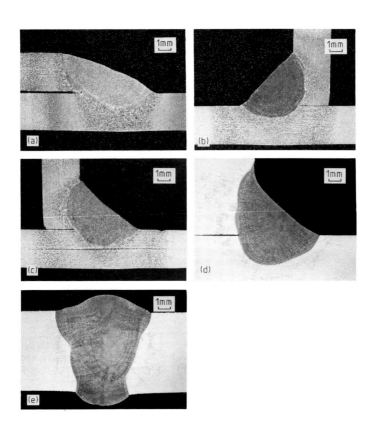

Weld	Method	Joint type	Material	Thickness, mm	Welding position	Welding current, A	Traverse speed, mm/min
a	Manual	Lap	Mild steel	3.0	Flat	200	180
b	Manual	Fillet	Mild steel	3.0	Flat	200	145
c	Manual	Fillet	Mild steel	3.0	Horizontal-vertical	200	175
d	Mechanised	Fillet	Austenitic stainless steel	6.0	Horizontal-vertical	210	130
e	Mechanised	Butt*	Austenitic stainless steel	6.0	Flat	Pass 1 210	180
						Pass 2 210	130

*60° V preparation, 1.0 mm root face, 1.5 mm root gap

35 *Sheet material welded with manual and mechanised hot wire TIG.*

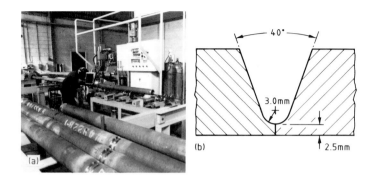

36 *Hot wire TIG welding of cast reformer catalyst tubes: a) General view of welding equipment; b) Joint preparation (courtesy of APV Paramount Ltd).*

Welding conditions					
Arc current, A	Arc voltage, V	Travelling speed, cm/min	Filler wire current, A	Deposition rate, g/min	Heat input, kJ/cm
280–320	10–11	8–13	120	30–37	22.6–23.7

70

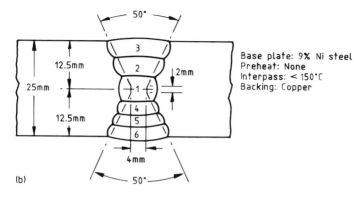

Base plate: 9% Ni steel
Preheat: None
Interpass: < 150°C
Backing: Copper

37 *Hot wire TIG welding of 9% nickel steel plate in the vertical position:
a) Welding equipment; b) Joint preparation and welding conditions (courtesy
of Kobe Steel Company).*

38 *Mechanised welding of a 180mm OD x 54mm wall C-pipe, internally clad
with 2.4mm of austenitic stainless steel, to a stainless steel flange. Runs 1 and
2 – austenitic stainless steel filler wire, remainder – Inconel filler wire
(courtesy of Equipos Nucleares SA).*

39 Four head welding machine for linepipe girth welding: a) General arrangement; b) Welding head (courtesy of Saipem SpA).

2 The DC current deflects the arc forward which facilitates an increase in welding speed and good sidewall fusion.

The technique has been used for welding plate up to 25mm (1in) in thickness in the fabrication of LNG storage tanks, for example, 9%Ni steels, where good mechanical properties are required; the joint preparation and welding parameters are also given in Fig. 37.

Hot wire (TIG) has also been used for welding thicker section material for example internally clad stainless steel pipe, 180mm (7in) OD x 54mm (2in) wall thickness, to a flange using a combination of austenitic stainless steel and Inconel filler, Fig. 38. The process has also been used in specialised systems for pipeline girth welds using a multi-head (four heads) to weld each quadrant of the pipe vertically downwards, Fig. 39.

Further reading

1   Mann A F: 'Hot wire welding and surfacing techniques'. *Weld Res Bull* 1977 223.

2   Harris I D: 'TIG hot wire offers high quality high deposition'. *Metal Construction* 1986 18 8.

3   Ogata Y and Aida I: 'A study on the improvement of TIG arc welding efficiency in out of position welding'. Kobe Steel Engineering Report No. 39 (April), 1980.

CHAPTER 7

Narrow gap TIG welding

Welding techniques

The narrow gap, or parallel sided, joint configuration offers a potentially more economic joining technique, because:

1 The amount of weld metal to be deposited is reduced;

2 Fewer passes are required to fill the joint.

Additional advantages are derived from the narrow weld and heat affected zone which, compared with the alternative V joint, produce lower residual stresses and distortion, and often superior mechanical properties. The reduction in the amount of weld metal required to fill the joint, compared with the normal V joint, is shown in Fig. 40a. Thus, in production, using the (narrow gap) technique a joint can be completed in a shorter welding time than with the conventional submerged-arc process, which has a vastly superior deposition rate, Fig. 40b. In the example of welding thick walled tubular components (illustrated in Fig. 40b), despite a deposition rate of 2 kg/hr (4.41 lb/hr) (arc time) compared with 6 kg/hr (13.2 lb/hr) for submerged-arc welding, the time to complete the joint by TIG was significantly less, at least for section thicknesses up to approximately 50mm.

In considering the use of the narrow gap technique, greater attention should be paid to machining so as to achieve closer fit-up tolerances. In addition, there is a greater need for tracking the electrode along the centre of the joint, as slight deviation from the centreline can result in lack of fusion defects because the arc and weld pool are attracted to the sidewall.

Equipment

For material thicknesses up to 25mm (1in), a parallel sided or slightly tapered joint preparation can be used with a joint gap of typically 6–8mm (1/4 to 5/16in). A conventional torch is used but with an extended electrode stickout. Process techniques such as current pulsing, use of argon-hydrogen

73

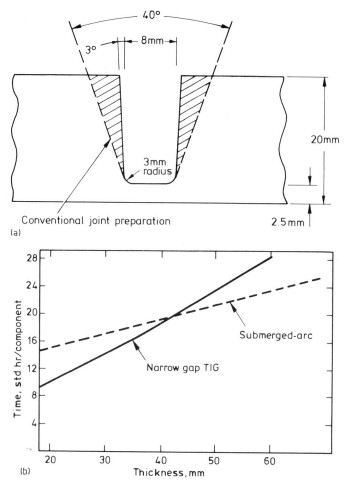

40 *Process features of the TIG-narrow gap techniques: a) Typical edge preparation used in the TIG narrow gap technique showing the reduced amount of filler metal required to fill the joint compared with a conventional V preparation; b) Time to weld tubular components of various wall thicknesses using the TIG narrow gap and submerged-arc welding processes.*

or helium shielding gas and arc oscillation have been employed to ensure a satisfactory weld bead profile. For arc oscillation, as the narrow joint generally precludes the use of torch weaving equipment, the arc itself is normally oscillated by an electromagnetic probe. However, as the depth to which the magnetic field can influence the arc is limited, this technique is normally applied only in wall thicknesses of less than 25mm (1in).

74

41 *Narrow gap welding equipment: a) Welding in a narrow gap joint using conventional (TIG-hot wire) equipment; b) Schematic diagram of welding operation using special purpose welding head; c) Welding head operating within a narrow gap; d) Production welding station (courtesy of Babcock Energy Ltd).*

The hot wire TIG process has also been applied as a means of increasing the deposition rate as illustrated in Fig. 41a.

The narrow gap welding operation is almost exclusively mechanised because of the need for precise positioning of the electrode in the centre of the joint. In the absence of suitable control devices, joint tracking is under the control of the operator. However, automatic arc length systems are almost always used.

At greater material thicknesses, typically up to 40mm (but thicknesses up to 72mm (3in) have been reported), the joint preparation is normally slightly tapered, to accommodate the closing of the joint during welding. In addition, because of the greater joint depth, specially designed torches must be used. Particular features of a narrow gap torch include dual gas shielding at the front and rear of the electrode, secondary gas shield from the top of the joint and water cooling down to the electrode tip to minimise electrode wear: the operation of the system is shown in Fig. 41b, a production torch operating in a narrow gap, Fig. 41c, and a production welding station in Fig. 41d.

Video monitoring equipment can greatly facilitate control of the position of the arc and the wire feed relative to the weld pool. An arc length control unit is also considered to be essential.

Applications

The technique has been used to advantage in the chemical industry, for welding 17.5mm (0.75in) wall thickness, 25%Cr, 20%Ni, 0.4%C (HK 40) tube. The joint was completed in only ten passes, as shown in Fig. 42. If the conventional V preparation had been used, approximately 20 passes would have been required.

The technique has also been applied in the power generation industry for welding pipes up to 700mm (12in) diameter and wall thicknesses up to 75mm (3in) in both ferritic and stainless steels, including transition joints. Although most reported applications are in the 1G position, specialised equipment has been produced for orbital welding of thick wall, low alloy steel piping, Fig. 43.

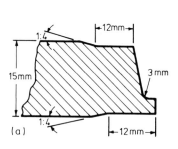

42 Narrow gap weld in 15mm (0.6in) wall thickness cast HK40 pipe using TIG-cold wire: a) Joint preparation; b) Section through weld (courtesy of APV Paramount Ltd).

76

43 *Narrow gap TIG welding head for orbital (5G) welding thick wall low alloy steam piping.*

Table 11 Examples of narrow gap/TIG-hot wire welding procedures for 50 mm thickness stainless steel (courtesy of Urantani et al)

Welding position	Joint preparation dimensions in mm	Number of passes	Welding parameters (3)							
			Welding current (1) I_P A	I_B A	Pulse condition T_P sec	T_B sec	Welding voltage, V	Welding speed, mm/min	Wire heating condition, A×V (2)	Wire feed rate, mm/min
Horizontal		1	160	120	0.3	0.5	10	90	–	500
		2	280	200	0.5	0.7	12	130	–	900
		Remainder	220	160	0.4	0.4	10	110	90–120A (2–4V)	800
			380	330	0.6	0.6	12	150		1800
Vertical		1	200	70	0.3	0.3	11	60	–	500
		2	280	100	0.5	0.6	12	100	–	900
		Remainder	220	160	0.4	0.4	11	80	90–120V (2–4V)	800
			380	230	0.6	0.6	12	130		1800

Note(1)

(2) Heating length: 40–60 mm
(3) Argon-helium shielding gas

The narrow gap joint configuration in combination with the TIG-hot wire process, can be used to increase further the joint completion rate. The standard commercially available equipment is suitable for wall thicknesses up to 25mm (1in) as shown in welding heavy walled pipe work, Fig. 41a, but as described above, at greater thicknesses special welding heads are normally required. The hot wire technique has been used in fabrication of 50mm thick stainless steel pressure vessels in the horizontal and vertical positions and in this case, the joint gap was 9mm wide; the joint configuration is given in Table 11 together with welding parameters.

Further reading

1 Hutt G A, Ward A, Cox B, Gough P C and Render G: 'Narrow gap welding'. *Metal Construction* 1984 16 6.

2 Render G S: 'Welding advances in power plant construction (Pt3)'. *Metal Construction* 1984 16 11.

3 Lockhead J C: 'Narrow gap welding'. South African Institution of Mechanical Engineers, 1 Paper 5, 1983.

4 Dawson D W O, Fivash R J and Ward A R: 'Automatic TIG welding'. *Metal Construction* 1976 18 3.

5 Hill R and Graham M R: 'Narrow gap orbital welding'. Welding Institute Conference, 'Advances in welding processes', Harrogate, May, 1978.

6 Harris L: 'Fabricating core barrels for nuclear reactors'. *Weld Des and Fab* 1980 6.

7 Urantani Y et al: 'Application of narrow gap GTA welding process to the welding of large type stainless steel pressure vessels'. IIW Doc XII-13- 83, IIW Doc XII-E-42-83.

8 Kripstrom K E and Pekkri B: 'TIG narrow gap orbital welding'. *Svetsaren* 1988 2.

9 Matsuda F et al (ed): 'Narrow gap welding (NGW), the state-of-the-art in Japan'. Tokyo 101, Japanese Welding Society, Japan, 1986.

CHAPTER 8

Process fundamentals

DC plasma welding

The similarities between TIG and plasma welding are readily apparent in that the arc is formed between a tungsten electrode and the workpiece but the torch arrangement generates the unique operating characteristics of the plasma arc torch. The electrode is positioned within the body of the torch, and the plasma forming gas is separated from the shielding gas envelope (Fig. 44). Thus, the emanating plasma is constricted by a fine bore copper nozzle. The most significant effect of plasma constriction is that the arc becomes very directional with deep penetration characteristics.

As shown in Fig. 45b the plasma arc assumes a columnar form compared with the conical TIG arc (Fig. 45a) at the same current.

The penetration capacity of the arc is determined by the degree of constriction of the plasma (diameter and length of the bore of the nozzle)

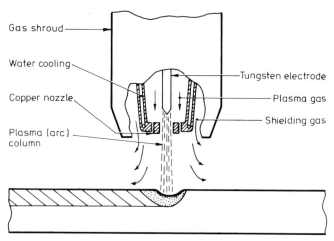

44 *Torch configuration for plasma welding. Note, electrode position and the addition of a separate plasma gas compared with the TIG torch (Fig. 1).*

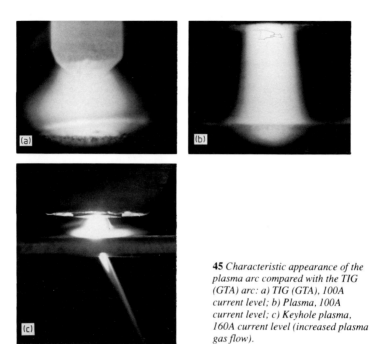

45 *Characteristic appearance of the plasma arc compared with the TIG (GTA) arc: a) TIG (GTA), 100A current level; b) Plasma, 100A current level; c) Keyhole plasma, 160A current level (increased plasma gas flow).*

and the plasma gas flow rate. The electrode angle has no effect on penetration and is usually maintained at 30°. However, as in TIG welding, the gas composition has a secondary influence on penetration. In this instance hydrogen, which increases the temperature of the arc by increasing the ionisation potential, as shown by the increase in arc voltage, is particularly effective. Helium is also used to increase the temperature of the plasma but, because of its lower mass, penetration can actually decrease in certain operating modes.

A particular feature of the plasma system is the pilot arc. Whilst the arc is again initiated by HF, it is first formed between the electrode and the plasma nozzle. Thus the pilot arc is retained within the body of the torch. When required for welding the pilot arc is transferred to the workpiece by completing the electrical circuit. Hence, the pilot arc system ensures reliable weld starting even under adverse conditions (such as long welding cables, well used electrodes and 'dirty' components).

Protection of the arc, weld pool and weld bead during solidification requires the use of a shielding gas as in TIG; the plasma gas alone is too turbulent because of the small nozzle to give adequate shielding.

The properties of the constricted plasma with variable arc force, which results from varying the plasma gas flow rate, have led to three distinct welding process variants:

– Micro-plasma welding: 0.1-15A;

– Medium current plasma welding: 15-100A;

– 'Keyhole' plasma welding: >100A.

Micro-plasma

Micro-plasma welding has been so termed because a very stable arc can be maintained , even at welding currents as low as 0.1A. It is possible to vary the arc length over a comparatively wide range, up to 20mm (0.75in) without adversely affecting stability and, because of the columnar nature of the plasma, without causing excessive spreading of the arc. With TIG welding, whilst the newer transistor controlled power sources can maintain an arc at currents as low as 1A (see Chapter 5). The arc is more sensitive to variation in torch distance, both with regard to stability and to spreading of the arc, because of its conical shape.

Medium current

At higher currents, that is up to 100A, the plasma arc is similar to the TIG arc, although it is slightly 'stiffer' and more tolerant to variation in arc length. The plasma gas flow rate can also be increased to give a slightly deeper penetrating weld pool, but with high flow rates there is a risk of shielding gas and air entrainment in the weld pool through excessive turbulence in the gas shield and agitation of the weld pool.

Keyhole

The most significant difference between TIG and plasma welding arcs lies in the keyhole technique. A combination of high welding currents and plasma gas flow rates forces the plasma jet to penetrate the material, forming a hole as in electron beam welding (Fig. 45c). During welding, this hole progressively cuts through the metal with the molten metal flowing behind to form the weld bead under surface tension forces, as shown in Fig. 46. The deeply penetrating plasma is capable of welding in a single pass,

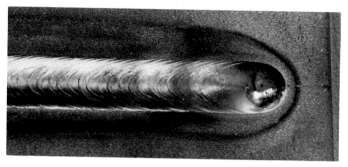

46 *Appearance of keyhole and solidified weld pool in welding 4mm thick, type 304 stainless steel.*

relatively thick sections within the range 3–6mm (0.1–0.25in). However, despite the tolerance of the plasma process to variation in torch to workpiece distance, this technique is more suitable to mechanised welding, as the welding parameters i.e. welding current, plasma gas flow rate and traverse rate, must be carefully balanced to maintain the stability of the keyhole and the weld pool. Instabilities can easily result in the loss of the keyhole giving only partial penetration of the weld bead and increasing the risk of porosity.

AC plasma welding

Sine wave

The AC plasma arc is not readily stabilised with sine wave AC for two reasons: arc re-ignition is difficult when operating with a constricted plasma and a long arc length; and the progressive balling of the electrode tip severely disturbs arc root stability. Thus, plasma welding of aluminium is not widely practised, although successful welding has been reported using DC (negative polarity) and helium shielding gas.

Square wave

The recent advent of the square wave power supplies described in Chapter 1 has made it possible to stabilise the AC plasma arc without the need for continuously applied HF for arc re-ignition. In addition, by operating with only 30% electrode positive the electrode is kept so cool that a pointed electrode tip and hence arc stability can be sustained. It is particularly important however, that to limit electrode/nozzle erosion the maximum current is reduced to less than that which can be operated with a DC (plasma) arc. For example, using 30% electrode positive, the current rating of a 4.8mm (3/16in) diameter tungsten electrode with a 40° tip angle would be reduced from 175A (DC) to approximately 100A (AC); the appearance of the electrode tip for 100, 120, 150 and 165A operation is shown in Fig. 47. Further, an increase in the proportion of electrode positive polarity, so as to improve arc cleaning, would significantly reduce the maximum operating current.

Despite the reduction in the maximum current at which the various electrode sizes can operate, stabilisation of the AC arc represents a significant advance in plasma welding. Until comparatively recently, when welding aluminium, no advantage could be taken of the deep penetration capability of the plasma arc because of the need to use a blunt electrode; the alternative AC TIG process produces shallow penetration. It is now possible to weld aluminium up to 6mm (0.25in) thickness in a single pass using the

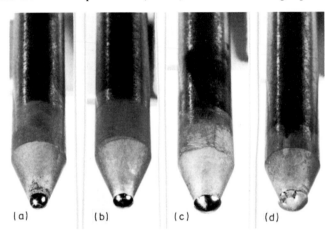

(a) (b) (c) (d)

47 *Appearance of electrode tip for various current levels. Electrode diameter, 4.8mm, initial tip angle 40°, 30% electrode positive polarity: a) 100A; b) 120A; c) 150A; d) 165A.*

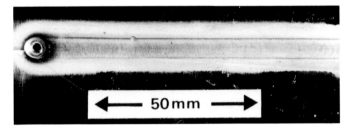

48 *4mm aluminium plate welded by the AC keyhole plasma process.*

keyhole mode. A butt weld is shown in Fig. 48. Because the arc scours the joint interface on passing through the material, very low weld metal porosity can be obtained.

Pulsed current (keyhole) welding

Similar benefits can be derived from pulsing the welding current in micro-plasma and medium current welding as described for TIG welding, but there are special advantages when operating with the keyhole mode.

The same principle applies, in that a high current pulse causes rapid penetration of the material and establishes a stable keyhole and weld pool. If this high current were maintained, the keyhole would continue to grow, causing excessive penetration and, ultimately, cutting would occur. Therefore, the pulse is terminated after a preset time and the weld pool allowed to solidify under a low background or pilot arc. It is equally important that the plasma gas flow be maintained during this period so that the keyhole does not close and, on re-applying the pulse current, the plasma can quickly penetrate the plate, re-establishing a stable keyhole and weld pool. Thus, welding progresses in a series of discrete steps with the pulse frequency balanced to the traverse rate to produce overlapping weld spots, as shown in Fig. 49.

Parameter selection

In pulsing the important variables are:

Welding current	Plasma gas
Pulse time	Pulse level
Pulse level	Background level
Background time	
Background level	

Selection of welding parameters can be simplified, first with the knowledge that the pulse time is determined more by the physical requirements of forming the keyhole and weld pool at a given traverse rate, than by the plate thickness or material composition. For most material within a plate thickness of 3–6mm (0.1–0.25in) a minimum pulse time of 0.1sec is required to re-establish the keyhole and weld pool. At greater pulse times, the excess energy is largely dissipated in the efflux plasma. The background time is usually set equal to the pulse time, which is sufficient for

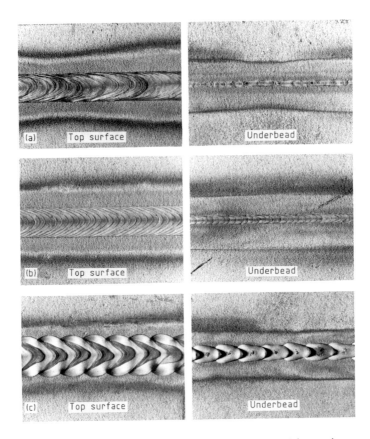

49 *Appearance of pulsed keyhole welds in 4.4mm austenitic stainless steel at different pulse frequencies. The background time is set equal to the pulse time: a) 8Hz; b) 2Hz; c) 0.5Hz.*

solidification between pulses. Thus, the pulse frequency is determined by the traverse rate and the need for at least 60% overlap of the pulses to provide a continuous seam. For instance, when welding 4.4mm (0.17in) stainless steel at 0.15 m/min (60 in/min) a suitable frequency is 2Hz. A pulse frequency of 8Hz gives insufficient time to re-form the keyhole weld pool, as shown by the intermittent penetration of the weld bead, whilst at a frequency of 0.5Hz, the weld spots become separated, giving pronounced undercutting on the top surface of the weld bead (Fig. 49(c)).

It follows that the pulsed current level and plasma gas flow rate are the major welding parameters which must be set to give an over-penetrating plasma for a particular material composition and plate thickness

Table 12 Welding parameters for keyhole plasma welding of austenitic stainless steel

Operating mode	Plate thickness mm	Plasma Bore, mm	Gas flow*, litre/min	Pulse frequency, Hz	Continuous current, A	Pulse current, A	Background current, A	Traverse rate, m/min
1 Continuous current	3.4	2.36	1.4	–	87	–	–	0.22
	4.4	2.36	1.75	–	120	–	–	0.22
	5.0	3.2	1.9	–	160	–	–	0.22
2 Pulsed current	3.4	2.36	1.4	2	–	115	20	0.15
	4.4	2.36	1.9	2	–	140	20	0.15
	5.0	3.2	2.3	2	–	190	20	0.15

*Plasma gas – argon, shielding gas – argon/5% hydrogen

combination. The background current is held low to give rapid cooling between pulses while the plasma flow rate is held constant to maintain the keyhole. For instance, when welding 4.4mm (0.17in) austenitic stainless steel, the pulsed current and plasma gas flow rate are typically 140A and 2 l/min (4 cfh) respectively. However, when welding the same steel in 5.0mm (0.2in) thickness, the pulsed current and plasma gas flow rate are increased to 190A and 2.3 l/min (4.9 cfh) and all other parameters are held constant. Typical welding parameters for stainless steel (at the sample pulse frequency and traverse rate) are given in Table 12.

The data in Table 12 are, of course, only a guide for the initial selection of welding parameters and must be used with caution, particularly when welding outside the 3–5mm (0.12–0.2in) thickness range. For example, in thinner plates pulse time must be reduced to avoid undercutting, whilst above this thickness range the pulse time is increased to avoid using excessively high currents. This might require a large nozzle bore and lead to correspondingly wide weld beads.

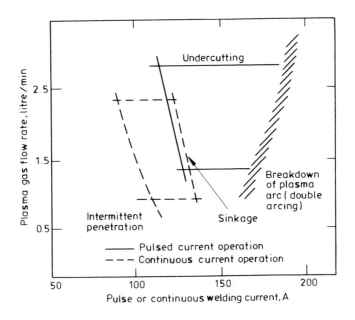

50 *Tolerance to variation in welding current and plasma gas flow rate in pulsed and continuous current keyhole welding. The boundaries show the welding parameter combinations at which specific defects are likely to occur.*

Pulsing the welding current overcomes one of the major difficulties encountered in keyhole operation, namely tolerance to variation in welding parameters. Using the simple acceptance criteria of full penetration, no undercutting and no bead sinkage, it can be shown that pulsing greatly increases the range of usable welding parameters.

As shown in the tolerance boxes for keyhole welding of 4.4mm (0.17in) thick stainless steel plate in Fig. 50, the range of acceptable pulse current is approximately twice that for continuous current operation. Consequently, the technique of alternate periods of melting and solidification produces a greater operating range or conversely, variations in the major welding parameters in pulsed operations are less likely to upset the process and result in defects such as lack of penetration, undercutting or weld bead sinkage.

Further reading

1   O'Brien R L: 'Arc plasma for joining, cutting and surfacing'. Welding Research Council Bulletin No. 131, July, 1968.

2   Omar A A and Lundin C D: 'Pulsed plasma – pulsed GTA – a study of the process variables'. *Weld J* 1979 58 4.

3   Bashenko V V and Sosnin N A: 'Optimisation of the plasma arc welding process'. *Weld J* 1988 67 10.

CHAPTER 9

Applying the plasma process

Practical considerations

The operation of the plasma process is essentially similar to TIG welding in that the arc is used as a heat source to fuse the joint and, when required, filler material is added separately in rod or wire form. In contrast to TIG welding, because the electrode is held within the torch body behind a small copper nozzle (Fig. 44), plasma welding has several singular operating characteristics:

1 A pilot (non-transferred) arc can be formed between the electrode and the copper nozzle; since a non-transferred arc is relatively inefficient as a heat source, an arc must be transferred from the nozzle to the workpiece for welding to enable heat to be generated in forming the arc roots.

2 The nozzle constricts the plasma to form a columnar shaped arc which, compared with the TIG arc, is more directional and less sensitive to variation in arc length; in TIG welding, because of its conical shape, the arc is more sensitive to arc length variation, both with regard to arc stability and the spread of the arc.

3 By increasing the plasma gas flow the penetration depth of the weld pool can be increased. In the keyhole mode the deeply penetrating arc plasma has sufficient power to cut completely through the material with the molten metal flowing behind to form the weld pool.

4 As the electrode is held within the torch body, i.e. behind the constricting nozzle, there is little risk of contamination from touching the weld pool or filler rod.

The practical operation of the plasma process is best considered in terms of the three distinct process variants noted before, which can be differentiated in terms of their operating current ranges.

- Micro-plasma welding, 0.1–15A

- Medium current plasma welding, 15–100A

- Keyhole plasma welding, >100A

The micro-plasma variant is applied in low current operations where the prime requirement is for a stable low current arc. The micro and medium current variants are normally employed manually, so that advantage can be gained from the tolerance to arc length variations, or in mechanised applications where the automatic arc starting and reduced electrode contamination features can be exploited. Because of the need for precise control of the welding conditions to maintain the keyhole, this high current variant can be applied only in mechanised operations.

Equipment

For micro-plasma operation the equipment is normally self-contained i.e. comprising a power source and control console. The torch is small and compact as shown in Fig. 51a.

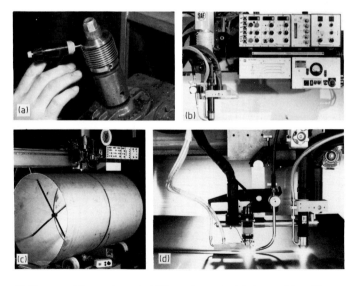

51 *Plasma welding equipment: a) Micro-plasma torch being used to weld stainless steel bellows (courtesy of BOC Ltd); b) Plasma arc welding head and control console; c) Industrial installation for large stainless steel vessels; d) Plasma-TIG (filler wire) torch equipment and TIG capping pass (b-d courtesy of SAF Welding Products Ltd).*

A typical plasma welding head and control system for medium and keyhole operation is shown in Fig. 51b. The plasma console is interfaced with a conventional TIG power source to control the pilot arc and the torch gas and water supplies. Special features of the system include welding current sequence and plasma gas slope-out controls. The latter feature is essential for closing the keyhole, for example when welding tubes.

An industrial installation for welding large diameter stainless steel vessels is shown in Fig. 51c.

The plasma arc can also be combined with a TIG arc to form (leading) plasma-TIG system as shown in Fig. 51d; welding speeds can be increased by up to 50% compared to a single torch system.

Electrode and nozzles

The electrode in the plasma system is normally tungsten-2% thoria. Typical electrode diameters, vertex angles and plasma nozzle bore diameter for the various current ranges are given in Table 13. At low and medium currents the electrode is sharpened to a point, whilst at high currents it is blunted to approximately 1mm diameter tip.

The plasma nozzle bore diameter, in particular, must be selected carefully, and it is prudent to use a nozzle whose current rating is well in excess of the operating current level. The plasma gas flow rate can also have a pronounced effect on the nozzle life with too low a flow rate possibly leading to excessive erosion. Multi-port nozzles, which contain two additional small orifices on each side of the main orifice, can be used at high current to improve control of arc shape. Use of an oval or elongated plasma arc has been found to be beneficial in high current welding, particularly when operating in the keyhole mode.

Plasma and shielding gas

Typical plasma and shielding gas compositions for the normal range of engineering materials are given in Table 14. The most common combination of gases currently employed in industry is argon for the plasma and argon or argon plus 2-5% H_2 for shielding. However, there are several other possible gases available which offer specific advantages.

Argon is the preferred plasma gas as it gives the lowest rate of electrode and nozzle erosion. Helium can be used for medium and high current operations to increase the temperature of the plasma which, in the melt (non-keyhole) mode, often promotes higher welding speeds. However, use of helium as the

92

Table 13 Maximum current for plasma welding for selected electrode diameter, vertex angle and nozzle base diameter (Levels are for guidance only, it is important to refer to manufacturer's recommended operating conditions for specific torch and plasma nozzle designs.)

Maximum current, A				Electrode diameter, mm	Vertex angle, degrees	Plasma*		Shielding†	
Torch rating, A						Nozzle bore dia, mm	Flow rate, litre/min	Shroud diameter, mm	Flow rate, litre/min
20	100	200	400						
Micro-plasma									
5				1.0	15	0.8	0.2	8	4–7
10						0.8	0.3		
20						1.0	0.5		
Medium current									
	30			2.4	30	0.79	0.47	12	4–7
	50					1.17	0.71		
	75					1.57	0.94		
	100					2.06	1.18		
		50		4.8	30	1.17	0.71	17	4–12
		100				1.57	0.94		
		160				2.36	1.42		
		200				3.20	1.65		
			180	3.2	60**	2.82	2.4	18	20–35
			200			2.82‡	2.5		
High current									
			250	4.8	60**	3.45‡	3.0		20–35
			300			3.45‡	3.5		
			350			3.96‡	4.1		

*Argon plasma gas   † Argon and argon-5% H_2 shielding gas   ** Electrode tip blunted to 1mm diameter   ‡ Multi-port nozzle

Table 14 Plasma and shielding gas compositions for plasma arc welding

Material	Plasma gas	Shielding gas
Mild steel	Argon	Argon Argon – 2–5%H_2*
Low alloy steels	Argon	Argon
Austenitic stainless steel	Argon	Argon – 2–5%H_2 Helium*
Nickel and nickel alloy	Argon	Argon Argon – 2–5%H_2*
Titanium	Argon	Argon 75% Helium – 25% Argon*
Copper and copper alloys	Argon	Argon 75% Helium – 25% Argon*

*Also used

plasma gas can reduce the current carrying capacity of the nozzle. Furthermore, because of its lower mass, weld pool penetration is reduced which, in certain materials, makes the formation of a keyhole difficult. For this reason, helium is seldom used for the plasma gas when operating with the keyhole mode.

Hydrogen is often added to the shielding gas, up to a maximum of 15%, to produce a hotter arc and a slightly reducing atmosphere. Hydrogen also constricts the arc which can increase the depth of weld pool penetration and promote higher welding speeds.

Helium, or a helium-argon mixture, typically 75% helium-25% argon, can also be used as the shielding gas. Whilst a hotter arc is generated it is less constricted, which can result in a wider weld bead compared with argon or argon-hydrogen shielding.

Joint preparation

As the selection of suitable joint preparations is largely determined by the type of component and the material thickness, specific joint designs are discussed in the appropriate sections on applications, but some general points are noted here.

Table 15 Joint configurations for plasma welding sheet and tubular components; for medium current plasma operating mode see also Fig. 7 (sheet) and Fig. 14 (tube).)

Thickness range, mm	Joint type	Joint configuration	Process variant	No. of runs	Comments
0.5–1.0	Micro lap	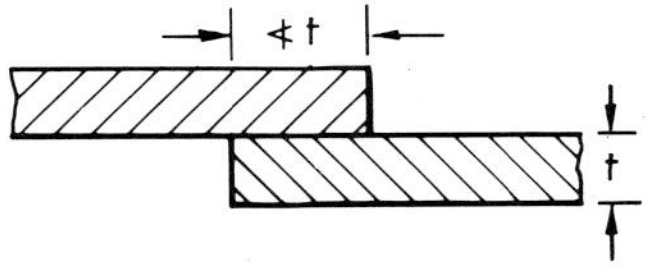	Micro plasma	1	Edges fully fused to produce additional weld metal – good clamping essential
0.5–1.5	Flanged edge	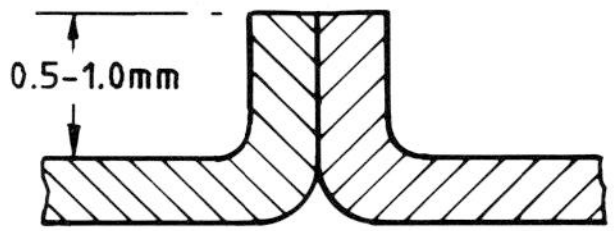	Micro plasma	1	Edges fully fused to produce additional weld metal
3.0–6.0	Square butt	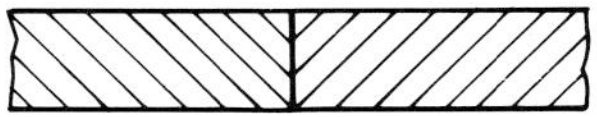	Keyhole plasma	1	Grooved backing bar required to prevent disturbance of the efflux plasma. Additional (cosmetic) run using melt mode may be employed
6.0–15	Single V butt	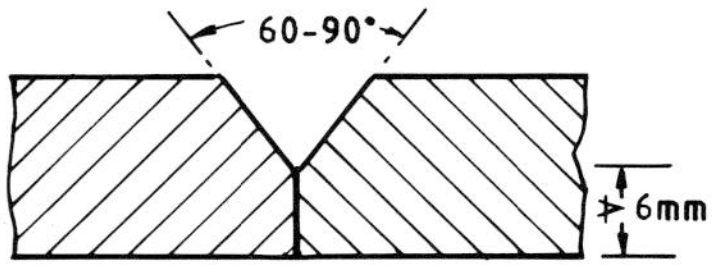	Keyhole plasma	2 or more	Keyhole technique used for root run only. Joint completed with the melt mode plus filler wire

95

In welding thin sheet material using the micro plasma operating mode, those designs incorporating integral filler material, see Fig. 28 and Table 15, provide some tolerance to joint fit-up and reduce the risk of burn-through. Use of a copper backing bar and finger clamps is strongly recommended when welding butt joints; to ensure a uniform heat sink.

As medium current plasma is employed as an alternative to TIG, the joint designs described in Chapter 2 can be used when the technique is applied to welding butt, T, edge and corner joints; typical sheet and tube edge preparations are shown in Fig. 7 and 14 respectively.

As the keyhole process variant has a deeply penetrating arc and weld pool, a greater sheet thickness, compared with TIG or the medium current plasma process, can be welded before an edge preparation needs to be employed (Table 15). It is current practice to limit the square edge closed butt joint preparation to 6mm (0.25in). Above this thickness, the normal V edge preparation is adopted, typically 60° included angle, with a root face of no more than 6mm.

Backing systems

The normal range of backing bar designs or shielding gas techniques, as previously described for TIG, is used when welding sheet by the micro and medium current techniques.

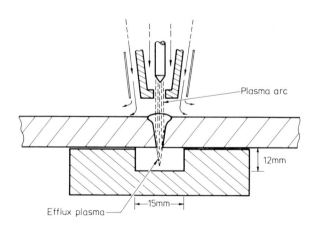

52 *Backing bar used in plasma (keyhole) welding.*

When applying the keyhole mode a grooved backing bar, with or without gas shielding, or total gas shielding of the underside of the joint must be used. Because the efflux plasma normally extends around 10mm (0.5in) below the back face of the joint, the groove must have sufficient depth to avoid any disturbance of the arc jet. Inadequate clearance gives the risk of turbulence in the efflux plasma arc, which disturbs the weld pool, causing porosity; a typical backing bar for plasma (keyhole) welding is shown in Fig. 52.

Industrial applications

Applications of plasma welding are considered in terms of the three process variants – micro-plasma, medium current and keyhole – and when describing specific applications, comparison has been drawn with the operating features of TIG welding to show why plasma was selected in preference to TIG.

Micro-plasma

The micro-plasma technique is particularly suited to welding sheet down to 0.1mm (0.004in) thickness, and wire and mesh sections. The narrow 'needle-like' stiff arc at welding currents within the range 0.1-15A prevents arc wander and minimises distortion; the equivalent TIG arc at this current suffers from instabilities because of arc wander, and is much more diffuse. Plasma can also be readily used manually as the torch is compact and there is high tolerance to torch-to-workpiece variation.

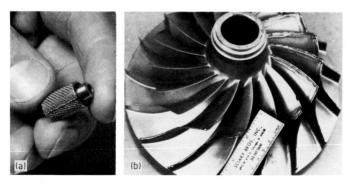

53 *Micro-plasma welding applications: a) Filter assembly (courtesy of BOC Ltd); b) APU impeller welded with pulsed wire micro-plasma process (courtesy of Huntingdon Fusion Techniques Ltd/Sciaky Bros Inc).*

Examples of the application of micro-plasma include thin section bellows (Fig. 51a), filter assemblies for the aerospace industry (Fig. 53a), and on-site welding of 0.3 and 0.5mm sheet for insulating elements in the advanced gas-cooled reactor (AGR) vessel. In the first two applications emphasis was placed on the need for careful consideration of component jigging, preparation of the joint edges and when possible, the use of a joint design which incorporates an integral filler to minimise the risk of burn-through. In the installation of the AGR insulating elements in areas of difficult access, the advantageous operating features of arc stability at low current levels, lighting of the joint area with the pilot arc before welding and tolerance to arc length variations were particularly important.

Micro-plasma with pulsed wire feed can also be used for building up the edges of worn impeller blades as shown in Fig. 53b. The impeller was manufactured in titanium 64 and edges were built up by 0.75mm (0.030in) in three passes using 0.875mm (0.035in) diameter filler wire.

Medium current

Plasma welding in the intermediate 15-100A range, and with the melt or keyhole operating mode, is more directly in competition with TIG. Successful applications have exploited the capacity to improve the depth-to-width ratio of the weld bead by use of higher plasma gas flow, and the position of the electrode within the body of the torch which can significantly reduce electrode contamination, which is particularly important in welding oily sheet material. These advantages, however, must be balanced against the increased bulkiness of the torch, which to some extent negates the advantages of the tolerance to variation in torch-to-workpiece distance in manual welding.

Examples of the manual use of plasma are found in the auto and chemical industries. In the fabrication of car bodies in approximately 1mm (0.04in) thick mild steel, plasma (braze) is applied to fill the joints between the panels to give a smooth surface after mechanical dressing; of other welding processes, MIG produces an unacceptable surface finish, whilst TIG suffers from difficulties in arc starting, the need to hold a short arc length and excessive electrode contamination from the surface oil.

Plasma has also been selected in preference to TIG for welding stainless steel pipes in fabricating brewery plant, which is particularly noteworthy as welding was carried out both in the factory and on site (Fig. 54a). The advantages claimed are better control of the weld bead penetration profile, which was particularly demanding in view of the stringent requirements of

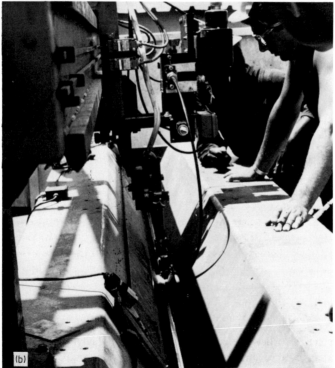

54 *Medium current applications of plasma: a) Manual welding of stainless steel pipes; b) Mechanised seam welding machine for fabricating large panels in 0.9mm thick stainless steel sheets (courtesy of William Press Ltd).*

crevice-free welds and maximum allowable pipe misalignment of ±0.13mm (0.005in) and the capacity to vary weld penetration by means of the plasma gas flow in preference to changing the welding current. The latter was found to aid positional welding and to facilitate control of the width of the weld pool.

Medium current plasma is also applied in mechanised operations and here the advantages of automatic arc starting and reduced electrode wear are particularly important. However, there is a need for close control of the welding parameters, particularly the plasma gas flow rate, and for regular equipment maintenance, to achieve consistent weld quality; a variation of 0.2 l/min (0.4 cfh) in the plasma gas flow rate significantly influences weld pool penetration. A unique application of mechanised techniques was welding 0.9mm (0.036in) thick stainless steel linings (1.6km (1 mile) in total) for a solvent copper extraction plant, which was installed in Zambia. The special plasma welding equipment adapted for welding 0.9mm (0.036in) sheet on site is shown in Fig 54b.

Keyhole plasma

As the keyhole operating mode has several special advantages (deep penetration, high welding speed) it is not normally in direct competition with TIG. The deep penetration capability, in particular, enables plate material up to 6mm to be welded in a single run or up to 12mm (0.5in) in two runs; the advantage over TIG welding is shown diagrammatically in Fig. 55 where 6mm (0.25in) wall thickness stainless steel tube which required four runs using TIG was welded in two runs using plasma. The most common technique is to carry out the first run using the keyhole mode and the second run using the melt mode if necessary with a filler wire addition; the practical arrangement for the plasma and TIG plus filler wire torches was shown in Fig. 51d. The limitations of the keyhole mode lie in the need for close control of welding parameters to maintain the keyhole and the difficulty of feeding filler wire into the keyhole without disturbing its stability. However, in sensitive materials, such as cupro-nickel alloys, the addition of filler wire during the keyhole run is essential to prevent porosity.

For these reasons the process is best applied to welding long linear seams in plate or circumferential joints in tube, in materials which are not susceptible to cracking or porosity in the autogenous welded condition. Orbital welding of tubular components is not usually practised because of the need to modify the plasma parameters to circumvent heat build-up to accommodate the changes in welding position as welding progresses around the joint, and

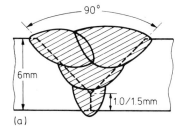
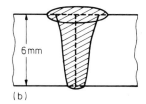
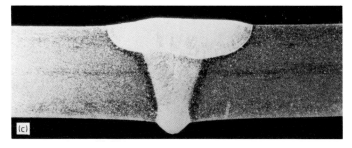

55 *Welding sequence in TIG and plasma welding of tubes; whilst the TIG weld requires four runs, the plasma weld is completed in two: a) TIG weld; b) Plasma weld, the first run is completed with the keyhole mode and a cosmetic pass is used to ensure a smooth surface profile; c) Typical appearance of two pass (keyhole plasma plus TIG) weld.*

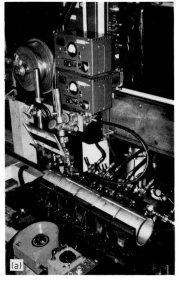

56 *Machine for plasma (keyhole) seam welding of pipe: a) Torch mounting and pipe clamping arrangement; b) Section through autogenous weld in 6mm thick stainless steel (courtesy of Devtec Ltd).*

to fill the keyhole on completion of the weld. Simultaneous sloping-out of the welding current and plasma gas can be employed to fill the keyhole but this requires specialised equipment and close control of the parameters to produce a satisfactory bead profile and to avoid porosity.

For the equipment shown in Fig. 51a, the manufacturer's suggested welding parameters for butt joints in stainless steel sheet are given in Table 16.

The keyhole technique has been applied successfully for several years in seam welding tube material in a range of thicknesses (6-12mm (0.25-0.5in)) and material compositions (stainless steel, Monel 400, and titanium). The welding equipment is shown in Fig. 56a and a section through the weld in Fig. 56b. Practical experience has shown that thicknesses up to 6mm can be

Table 16 Welding parameters for plasma (keyhole) welding of butt joints in stainless steel (courtesy of SAF)

| | | | Filler metal | | Gases used (flow rate, l/min) | | |
Thickness mm	Arc current amperage, A	Welding speed, cm/min	φ, mm	Wire feed speed, cm/min	Plasma producing gas argon	Nozzle argon 2/5%H$_2$	Additional argon
Stainless steel							
2	120	65	1	60	2–3	15	–
	Pulsed current						
3	130–140	45–50	1	50	3–4	20	–
4	150–160	35–38	1.2	60	4-5	20	–
5	150–160	28–32	1.2	60	4.5–6	20	–
6	160–180	26–32	1.2	60	8–9	25	–
8	250–280	18–20	1.2	90	8–10	25	–
Titanium							
6	220	26	–	–	7	30	30
	180	15	–	–	2	30	30
8	240	26	–	–	8	30	30
	180	15	–	–	2	30	30
10	940	20	–	–	10	30	30
	240	20	–	–	2	30	30
Zirconium							
5.8	140	98	–	–	3	30	40
7.2	150	97	–	–	4	95	40
	140	19	–	–	2	20	40

Shielding on underside of weld–argon or argon–H$_2$ (10–90 l/min)

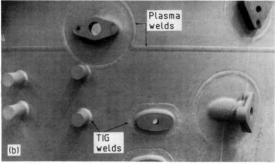

57 *Plasma welding aero engine components: a) Intermediate compressor casing for the RB 211 jet engine (the complete casing held in the positioner); b) Circumferential and boss welds completed by plasma and TIG (courtesy of Rolls-Royce Plc). Typical welding parameters for plasma welding: plasma and shielding gas composition – argon, plasma nozzle orifice diameter – 2.35mm, plasma gas flow rate – 1.2 l/min, welding current – 55A welding speed – 0.15 m/min.*

welded in a single pass, but greater thicknesses require the use of joint preparations of 75° (included angle) and a 5mm (0.2in) root face. The weld is then completed in two passes – an autogenous plasma (keyhole) root run followed by a capping run of plasma (melt mode), or TIG plus filler wire.

An equally successful application has been in the fabrication of high integrity components for the aerospace industry. A notable example is that of the intermediate compressor casing for the Rolls-Royce RB211 jet engine, Fig. 57a. The appearance of the circumferential weld and the smaller boss insertions in the casing wall are shown in more detail in Fig. 57b; also shown are manual TIG welds which, because of their non-circular

58 *Plasma keyhole welding of titanium showing the use of a glass cap to give total shielding of the weld and surrounding area (courtesy of Rolls Royce Plc)*

profile, would be difficult to weld using the keyhole process. As the material thickness in this instance is only 2.6mm (0.1in) preference for the plasma process is based solely on the integrity of the weld. The narrow weld and heat affected zone results in sound welds and low distortion in materials such as Jethete M152 and 12%Cr air hardening material.

Plasma (keyhole) welding also has particular advantages for welding titanium alloys, producing very low porosity without resort to special preparation of the joint edges. This is directly attributable to the scouring of the arc as the keyhole cuts through the material; an example of plasma welding of a 2mm thick titanium casing in the aeroengine fabrication is shown in Fig. 58. The effectiveness of the plasma process in making a narrow weld which cools quickly means that only simple shielding of the weld bead is necessary in production. A trailing gas shield is sufficient, compared with the more common use of glove boxes or vacuum chambers for TIG welding titanium, particularly where the section thickness is greater than 1.5mm (0.06in).

Pulsed (keyhole) welding

The comments on pulsed current TIG apply equally to plasma welding for the melt type operating mode, but special advantages are to be gained from its adoption in the keyhole operating mode.

Despite the successful application of keyhole plasma described above, the technique is not widely applied in industry. The major reasons are:

59 *Sections through plasma keyhole welds in 4.4mm austenitic stainless steel, showing the effect of pulsing the welding current on weld bead penetration profile: a) Continuous current; b) Pulsed current.*

Typical welding parameters

	Continuous	*Pulsed*
Plasma gas composition	*Argon*	*Argon*
Plasma nozzle diameter, mm	*2.36*	*2.36*
Plasma gas flow rate, l/min	*1.75*	*1.9*
Plasma current, A	*120*	
Plasma current, pulsed, A		*140*
Pulsed frequency, Hz		*2*
Welding speed, m/min	*0.22*	*0.15*

1 Close control is required of the major welding parameters to maintain keyhole/weld pool stability.

2 The complex torch arrangement requires more than normal planned maintenance to ensure reproducible performance.

Defects which are frequently observed in practice include undercutting through too high a plasma force and 'humping' when welding at high traverse speeds. Catastrophic breakdown of keyhole/weld pool stability results either in partial penetration of the plate and excessive porosity in the weld bead, or in cutting without establishing the weld pool. Consequently precautions must be taken to avoid destroying the component and incurring expensive repairs.

60 *Pulsed keyhole plasma weld in 40mm OD, 4mm wall thickness stainless steel steel pipe, welded in the 2G (pipe rotated) position: a) Surface appearance; b) Section through weld. Typical welding parameters: plasma gas composition – argon, plasma nozzle diameter – 2.36mm, plasma gas flow rate – 1.9 l/min, plasma current – 125A, Pulsed frequency – 2Hz, welding speed – 0.15 m/min*

Pulsing the welding current overcomes the major limitation of poor tolerance to variation in welding parameters: as shown in Fig. 50, tolerance to variation in welding current and plasma gas flow is greatly increased through the intermittent solidification of the weld pool. Improvements have also been observed in the weld bead penetration profile, as shown in Fig. 59. The continuous current weld has the characteristic 'wine glass' penetration profile, which is particularly narrow in the centre of the plate. The underbead in materials such as stainless steel is invariably 'peaky' with a sharp angle of contact with the surface of the parent plate (Fig. 59a). In contrast, the penetration profile of the pulsed keyhole weld has a more uniform width through the thickness of the plate, and has a flatter underbead (Fig. 59b). Thus, the increased width of the weld bead through pulsing of the current is advantageous in reducing the demands on joint tracking.

In application of the technique to welding pipe it has been demonstrated that stainless steel pipes of 40mm (1.6in) OD, 4mm (0.16in) wall thickness, can be welded in the torch horizontal and pipe vertical position (rotated pipe) without an edge preparation. The surface appearance of the horizontal weld is shown in Fig. 60a, and a section through the weld in Fig. 60b. The section should be compared with that of the typical pulsed TIG welded pipe of 5mm (0.2in) wall thickness shown in Fig. 22, where the weld required an expensive U type joint preparation (Table 7) and five runs to complete the weld. The specific problem in pipe welding, that of closing the keyhole on

106

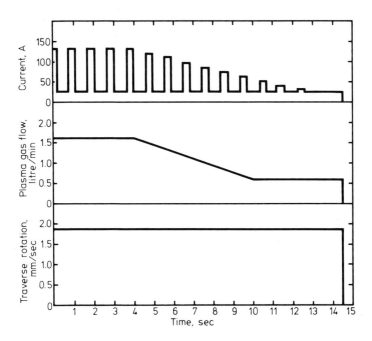

61 *Sequence of operations required to fill the keyhole on completion of orbital weld. Note, simultaneous sloping-out of the welding current and plasma gas flow.*

completion of the operation, was overcome by simultaneously sloping-out the welding current and plasma gas flow, the sequence of operations is given in Fig. 61.

Although the pulsed plasma process has not been widely adopted in industry, applications have been reported in the US and more recently in the UK for the manufacture of aeroengine components. In the US, 25mm (1in) diameter, 3.1– 4.6mm (0.12–0.18in) wall pipes in $2^{1}/_{4}$Cr– 1Mo steel were fabricated for the steam generators of the high temperature gas-cooled reactor. The advantages of the plasma arc process compared with the alternative TIG were higher welding speeds, greater depth of penetration, simpler joint preparation for a weld quality matching that observed in TIG welding.

In the manufacture of aeroengine components, 50mm (2in) diameter bosses were welded into a titanium casing which had a wall thickness of only 2mm (0.08in). Pulsing was carried out purely to enable the keyhole mode to be applied in such a section thickness and to produce a satisfactory surface appearance, particularly in the slope-out region. The reason for employing the keyhole mode was to achieve low porosity (from the scouring action of the arc as it passed through the material section) and to minimise distortion.

Further reading

1 Ford E K: 'Plasma and TIG tube production'. TIG and plasma welding, The Welding Institute, 1978.

2 Haylett R: 'Microplasma welding – some applications'. TIG and plasma welding, The Welding Institute, 1978.

3 Lea N H: 'Plasma welding – stringent specifications, successful applications'. TIG and plasma welding, The Welding Institute, 1978.

4 Norrish J: 'Recent applications in plasma welding'. TIG and plasma welding, The Welding Institute, 1978.

5 Holko K H: 'Plasma arc welding $2^1/_2$Cr-1Mn tubing'. *Weld J* 1978 57 5.

6 Numes Jnr A, Bayless Jnr E, Jones III C, Munsfo P, Biddie A and Wilson W: 'Variable polarity PAW on space shuttle external tanks'. *Weld J* 1984 63 9.

7 Craig E: 'The plasma arc process – a review'. *Weld J* 1988 67 2.

8 Tomsic M and Barhorst S: 'Keyhole plasma arc welding of aluminium with variable polarity power'. *Weld J* 1984 63 2.

9 Woolcock A and Ruck R J: 'Keyhole plasma arc welding of titanium plate'. *Metal Construction* 1978 10 12.

10 Haylett R: 'Micro-plasma welding – some applications'. TIG and plasma welding, The Welding Institute, 1978.

CHAPTER 10
The future

The TIG and plasma processes are used where high quality welds must be achieved, and although they are often considered to be in competition, specific advantageous features can make a particular variant the preferred choice on purely economic or quality considerations. Secondary factors such as available workforce skills, ease of use or even initial cost of welding plant can have a determining influence. It is hoped that the detailed information given here will lead not only to a more uniform choice but also to the advancement of process technology.

With regard to progress in TIG and plasma in the 1990s, the potential of transistor power sources and microcomputers deserves particular mention. The accuracy and flexibility of commercial transistor power sources will enable the processes to be used more widely and in more demanding conditions. This will be particularly true in mechanised operations where the operator has little or no control over the behaviour of the weld pool. The capacity to set and to maintain the welding parameters will remove one possible production variable. Furthermore, as the power sources lend themselves readily to programming and to the use of feedback systems for keeping constant the welding parameters and the weld bead penetration profile, transistor systems will play an increasingly significant role in controlling welding.

Microprocessor based systems, exploiting the advantages of flexibility in design, memory and computation will also promote the increased use of mechanised and automatic welding systems. For example, in the automation of the TIG and plasma processes, microcomputers will be capable of storing optimised welding parameter values, and hence suitable parameters, even for the more complex welding operations such as pulsed TIG, should be readily produced from the input of information on the material, joint type, welding position, etc. The development of these 'intelligent' devices will greatly enhance the performance capabilities of the techniques described, leading to more economical welding methods and more consistent weld quality.

INDEX

AC plasma   83
AC TIG   22
Arc plasma   9, 80
Argon-helium   57
Argon-hydrogen   14
Automatic wire feed   36

Back face   57, 59
Backing bar   26, 96
Backing systems   25, 27, 96
Basic function systems   49

Cast to cast variation   44
Clamps   37, 63
Cleaning   28, 56
Cost analysis   52

DC TIG   9
Defects   40, 41, 105
Deposition rates   65

EB insert   46
Edge preparation   96
Electrode
   angle   81
   composition   9
   diameter   11, 12, 62
   oscillation   49, 51, 92
   tip   11, 62
Equipment   47

Filler rod angles   29
Flow rate   17, 86
Front face machines   57

Gas
   backing systems   27, 40
   composition   81
   shield   76

Helium   16, 81, 92
Hot pass   49
Hot wire TIG   65, 79
Hydrogen   94

Inserts   46
Intermediate systems   49

Joint
   configurations   56, 57, 59
   fit-up   37, 62
   preparations   25, 38, 73, 94

Keyhole plasma   82, 100

Manual welding   29
Material variation   44
Mechanised welding   37
Medium current plasma   82, 98
Micro-plasma   82, 97
Micro-TIG   62

Narrow gap   73, 79
Negative purging   47
Nitrogen   16
Nozzle bore   92

Parameter settings   19, 31
  aluminium-magnesium   31
  butt joints   31, 102
  cupro-nickel   54, 55
  mild steel   32
  Monel 400   102
  stainless steel   20, 32, 102
  titanium   102
Pilot arc   81
Pipes   29
  stainless steel   51
Plasma   9, 80, 92
Power sources   12, 15
Pulsed current   18, 38, 39, 44, 47, 55, 61, 85, 86, 104
Pulsed wire feed   49

Ring insert   38
Root pass   49

Shielding gas   14, 16, 17, 44, 92
Sine wave arc   22, 83
Square wave arc   22, 83
Stainless steel   106

Thin sheet   39
Titanium   104
Torch angle   29
Tube welding   38, 47, 56
  carbon-manganese   57
  to tubeplate   56
Two operator technique   29

Undercutting   105

Vertex angles   12, 62, 92

Welding
  parameters   85
  systems   47
  techniques   29, 37

Welding International

Welding International provides translations of complete articles selected from major welding journals of the world including those from the USSR, Japan, China, Italy, Poland, Czechoslovakia and the German Democratic Republic.

Each issue contains about 85 pages of translated articles, including illustrations, covering research techniques, equipment and process developments, applications and materials. Translated contents pages from the latest issues of major welding journals worldwide are also published in *Welding International* and subscribers are invited to nominate articles they wish to see in English. When there is sufficient demand for a specific article, it is translated and published in the journal.

Welding International therefore provides a valuable and unique service for those needing to keep up-to-date on the latest developments in welding technology in non-English speaking countries.

12 issues per year.

For information on subscriptions contact:
Abington Publishing,
Abington Hall, Abington, Cambridge CB1 6AH
Tel. 0223 891358   Fax. 0223 893694